SCHRIFTEN DER WISSENSCHAFTLICHEN GESELLSCHAFT
AN DER JOHANN WOLFGANG GOETHE-UNIVERSITÄT
FRANKFURT AM MAIN

NR. 14

FRANZ STEINER VERLAG STUTTGART
2000

DIETER STEFAN PETERS
AND MICHAEL WEINGARTEN (EDS.)

ORGANISMS, GENES AND EVOLUTION

EVOLUTIONARY THEORY AT THE CROSSROADS

PROCEEDINGS OF THE 7TH INTERNATIONAL
SENCKENBERG CONFERENCE

FRANZ STEINER VERLAG STUTTGART
2000

Gedruckt mit finanzieller Unterstützung
der Stadt Frankfurt am Main

CIP-Titelaufnahme der Deutschen Bibliothek
Organisms, genes and evolution : evolutionary theory at the crossroads ;
proceedings of the 7th International Senckenberg Conference / Dieter
Stefan Peters and Michael Weingarten (ed.). – Stuttgart : Steiner,
2000
    (Schriften der Wissenschaftlichen Gesellschaft an der
    Johann Wolfgang Goethe-Universität Frankfurt am Main ;
    Nr. 14)
    ISBN 3-515-07659-X

ISO 9706

CONTENTS

PREFACE

The 7th International Senckenberg-Conference was held at the Forschungsinstitut Senckenberg in Frankfurt am Main, 9th – 12th October 1996.

When Wolfgang Friedrich Gutmann, the initiator of the conference welcomed the participants nobody could foresee that this meeting was to be the last event organized by him. Wolfgang Gutmann died suddenly in the morning of April 15th 1997. We dedicate this volume to our collaegue honoring a true friend and a most creative scientist whose impact on Vertebrate Morphology, concepts of the organism and on evolutionary thinking cannot be overrated.

The tragic circumstances of Professor Gutmann's death regrettably delayed the completion of the volume. We ask the contributors for their understanding and apologize for any inconvenience caused. Most unfortunately the manuscript of Professor Gutmann's own contribution to this symposium volume is lacking. The editors found only a sketchy draft of his contribution and felt incapable to develop the available notes into the original text presented by him at the conference.

Nevertheless the contributions of this volume reflect very precisely the varied results of the meeting outlining the broad arena of modern evolutionary biology expressed in the theme of the meeting: "Organisms, Genes and Evolution – Evolutionary Theory at the Crossroads". Evolutionary biology is still marked by a struggle for an adequate theoretical basis. The organismic versus the reductionistic approach seem to be the major conflicting alternatives. The discussion is far from a definite conclusion, although arguments in favour of the organismic view seem to gain strength and persuasiveness.

The editors express their appreciation to the Senckenbergische Naturforschende Gesellschaft for supporting the conference financially and morally, and to the Wissenschaftliche Gesellschaft an der Johann Wolfgang Goethe-Universität for making the publication possible. Many thanks also to Walter Bock and Peter Beurton for their editorial support.

D. Stefan Peters   Michael Weingarten

Where does biology get its objects from?

PETER JANICH

1. Introduction

On reading the abstracts of the "7. Internationales Symposion der Senckenbergischen Naturforschenden Gesellschaft" on "Organisms, Genes, and Evolution", the alert reader may find two things remarkable. First, the authors do not agree on a uniform use of their words, neither with respect to the biological terminology (organism, gene, species, evolution, etc.) nor with respect to the philosophical one (explanation, causation, teleology, holism, reduction, idea, etc.). And secondly, their disagreements and controversies do not in general focus on substantial biological questions but rather on *methodological and epistemological* ones. A brief diagnosis could therefore be stated thus: There are differences of philosophical opinion, and the opponents are talking at cross purposes in terminological respects.

Any efforts to improve on this situation by philosophical means will have to take a stand on the question what kind of *task* there is *for philosophy*. For explicatory purposes, we will oppose two fundamentally different approaches which, just as in the case of other scientific disciplines, reveal themselves in a particular attitude towards biology as it is actually practised. Programmatically, they can be sketched thus:

(1) On the one hand, there is a *descriptive philosophy* of science, borrowing from the historiography and sociology of science and/or interested in a systematical and structured exposition of the prevalent biological sub-disciplines as it is presented by the so-called analytical philosophy.

(2) On the other hand, there is a critical analysis of knowledge claims, combined with a methodical reconstruction of the historical-actual discipline of biology. This project includes explicitly *normative components*.

Although these differing kinds of philosophy do sometimes appear to be entrenched in an unending, pseudo-religious war, they can more profitably be viewed as potential collaborators in a division of labour. From that perspective, the results of a descriptive philosophy of biology can be put to use in posing and answering questions concerning the validity of knowledge claims, the methodological means for the formation of valid theories based on the different types of biological experience, and, related to this task, the procedures for the formation of concepts and theories. Thereby, a critical evaluation of the terminology of biology, its models and claims can be combined with a systematic clarification of its fundamental concepts.

2. The biologists' biology

As a research speciality and university subject, what we call "biology" is the product of its historical genesis. In this respect, it is comparable to a product of unintentional, natural evolution and, in virtue of this, merits epistemological reflection. Even if each individual scientist were to make his aims and means fully explicit, including a definition of his terminology, this would not suffice to confer the
same properties on biology as a discipline, since the aims, means and terminology
of such a discipline are never the product of an individual agent. Rather, the shared
language, the preferred theory, the most successful models, the biological "state of
the art" is always the complex result of a co-operative activity for which no single
scientist is fully responsible as an individual. Therefore, even on the counterfactual
assumption that each individual scientist were an optimally conscientious agent with
respect to the epistemological reflection of his work, there would still remain epistemological deficits regarding scientific disciplines on the whole.

Unfortunately, the assumption that scientists are disposed towards the epistemological reflection of their work seems to be ill-founded. This is indicated by a
simple but nevertheless consequential example. Usually, biologists use the words
"biological" and "biotic" interchangeably. However, on reflection it should be obvious that *"biological"* is derived from "biology", the *name of a scientific discipline*,
whereas *"biotic"* roughly designates the *subject-matter* of that discipline.

Of course there are cases in which this kind of terminological laxness does no
harm because the context leaves no doubt as to whether we are talking about the
discipline biology as an historically changeable complex of human activity, or about
organisms, their structure, their function, their heredity etc. Consistent neglect of
this distinction, however, reveals a lack of "epistemological reflection". This means
that by focusing on object level, biological questions, biologists tend to disregard
the fact that they are engaged in a human cultural endeavour constituted by their
own activities, their decisions regarding aims and methods of research, their implicit understanding of science, causality, explanation, nature, etc. This carries the danger of ascribing to the (biotic) subject matter of biology what in truth is due to a
decision of the scientist concerning his (biological) research activity.

Even worse, what tends to get lost is that *the objects of biology are not "given"*
as the animals and plants of an uninhabited area seem to be to the layman who, being
unable to discern any consequences of human intervention, considers them to be
part of an untouched, "natural" nature. By contrast, the objects of biological science
are funded on a number of preparatory, human decisions. Which questions and problems guide the description of living matter? Which criteria are established in order
to distinguish valid from invalid answers and solutions? Which procedures are heed-
ed in observation, experiment, explanation, or retrodiction? In short, the *objects of
the science of biology* are, from an epistemological point of view, the *products of
human actions* and, as such, artifacts subject to evaluation in terms of the rationality
of means and ends.

Consequently, to analyse the biologists' biology in epistemological terms means
not to part company with them when it comes to the cavalier obfuscation of the

distinction between "biological" and "biotic". Rather, what has to be asked is where do biologists get their objects from, which claims do they make on their behalf, and how can these be substantiated.

3. The descriptivist trap

Just as biology, *philosophy* is the product of a historical co-operative activity. In this sense, its results can be regarded as quasi-naturally grown, too. In the present, the predominant attitude takes the philosopher's task to be the *description and analysis* rather than the critique of knowledge. This is evident from the fact that the extant sciences in terms of their *published results are "analysed"*, where the focus of attention usually is on their most recent and, consequently, from the scientists' point of view, their most attractive theory. The description and analysis deals with the terminological meanings of *fundamental concepts* (such as organism, organ, species, gene, homology), with the syntactical structure of *theories* (e. g., the so-called "Synthetic Theory"), with the identification of *research programmes* (e. g., the reduction of biological explanations to molecular ones), and more. Characteristically, the *validity of the results* which are analysed is not questioned in the description and analysis but *remains an unproofed presupposition*. In other words, the results of such analysis rest on an implicit (or, perhaps, sometimes explicit) premise, according to which we learn something about biology because, and only because, the biological results themselves already constitute knowledge. Only the presumed success of biology makes it the worthy subject of analytical philosophy of science. This premise is also effective in the descriptions which are given of the particular activities scientists engage in, of the individual observation, the data basis, the fundamental distinctions etc. In other words, for the descriptivist it is the accepted theory with general sentences, principles, and laws, which explains the singular observation or description. And it is the acceptance of a theory which makes the methods by which it had been constructed appear to be justified *ex post* in virtue of their success.

To this kind of "top-down-philosophy" (i. e., a philosophy proceeding from the "top" of the most recent theories "downwards" to their empirical and conceptual basis and the relevant methods) we will here oppose a "bottom-up-philosophy" which starts from the context of pre- or extra-scientific practices and works its way "up" to the realm of scientific methods and theories. Thereby, we will steer clear of the presumption in favour of the product of an at least partly contingent historical process, namely, in favour of biology as we know it. (Neither will it be presumed, of course, that biology in its actual state is an altogether ill-conceived matter.) The *historically-factual opinions of scientists* will serve as a *point of departure* from which the investigation starts by asking for terminological explicitness and empirical content.

The point of this reversal is to reinstate the *basic epistemological question* which cannot be asked anymore if the *validity* of the biological results in hand is presumed from the start, namely, the "critical" – literally, the "distinguishing" – question how theories are to be evaluated with respect to their soundness and acceptability.

In asking this basic epistemological question, whether the theories advanced in public really constitute knowledge, the aim is to *refrain from any dogmatic presumption* both for or against the received opinion. Nevertheless, what is of interest is not a philosophical fiction but biology as it is actually practised by biologists. What has to be renounced is only the presupposition that what is factually being accepted is justified already because of this. The task then is to find and establish *new criteria of validity* other than the factual acceptance by the specialists.

4. Biology as a cultural accomplishment

Had it not been consistently disregarded, the following observation would sound annoyingly trivial: *Biology is made by people.* Even if, from a historian's point of view, it may seem to be something like the product of a natural growth process, the development of a scientific discipline *results from intentional actions* by individual agents, not from lawful events which happen to the scientists.

In this process, no scientific discipline is or can be isolated from a practice which encompasses the scientist's everyday life. The historical development of the science of biology is not independent from the lifeworlds [*Lebenswelten*] of the principal agents. Nor would it be reasonable to assume that someone could become a biologist unless he had at his disposal the wealth of abilities such as linguistic competence, intentional action, the systematic provoking of experiences, and everyday discriminatory capabilities, which are all part of our lifeworlds. Rather, these still constitute an important background for the biologist who describes, discriminates and investigates plants and animals.

This indicates the twofold dependence of the modern science of biology on a pre- and extra-scientific lifeworld: (1) Systematically, biology as done by human beings is embedded in a web of individual and social conditions and expectations (consequentially, it cannot exclusively refer to "nature", "natural laws", or any properties of the subject matter of biological research). (2) Historically, the science of biology has not been created from scratch by an act of volition but rather resulted from the incremental "scientification" of non-scientific practices. Even if these practices are in turn influenced by the products of highly advanced, modern sub-disciplines of biology such as genetic engineering, they still remain the sphere from which basic questions and problems originate and to which we turn for the justification of aims of research and the evaluation of results. For this reason, we now have to turn toward the question which practices are at the basis of biology, and how these are transformed and refined into a scientific discipline.

5. The object of biology

From the very beginning of human civilisation – one might even say: by definition so –, human beings have learned to *utilise plants and animals* for their survival. The domestication and breeding of plants and animals for purposes of food, garments, shelter, protection, and workforce have gradually been established as a collective, culture-constitutive capacity – certainly in parallel with the discrimination of and defence against poisonous, aggressive and dangerous plants and animals. It deserves to be mentioned that the word "culture" derives from the Latin "cultura", designating the agricultural practices. Thus, what nowadays mainly connotes the arts and sciences originally referred to the *human interference with nature* as we find it, resulting in its reshaping according to human purposes.

A second kind of archaic cultural accomplishment which is to become a practice fundamental for the future development of biology concerns the way man deals with himself, his fellow men, relatives, and enemies. The perspectival perception of self and other, which we nowadays assign to the subject matter of biology, medicine, and psychology, is intimately related to the human neediness concerning protection against life-threatening dangers, basic metabolic requirements, or health care, but also to the assessment of sexual attractivity, helpfulness as a collaborator, or powerfulness as either a comrade-in-arms or enemy.

Presumably, these two *practical spheres*, i. e., the utilisation of plants and animals on the one hand, the dealings with humans on the other, have increasingly *overlapped and intercalated*. Similarities between man and animals and their life processes will have been as conspicuous then as they are now in the realm of ordinary experience.

The "scientification" of such practices can be divided into three aspects and thereby represented as a step-by-step process:

(1) "*Object constitution*": Any interactions with plants and animals and their gradual transformation from natural, given things into humanly influenced ones (by cultivation, improvement, and reaping of plants, or by animal husbandry, breeding, and utilisation of animals) "constitutes" them *as objects of human perception and action*. It is not passive, even meditative observers of an untouched nature, but rather hunters, gatherers, farmers, butchers and warriors who arrange the flora and fauna of their environment for themselves and adopt it to their purposes by practically discriminating amongst its members. Irrespective of the fact that human manipulation cannot produce things at will but is confined to the empirically possible, plants and animals are being made cultural objects.

These pre-scientific ways of imposing a structure on the plant and animal world reappear in the different steps of "scientification". Here, we have to consider the epistemic aims of the palaeontologist, the taxonomist, the evolutionary theorist, the ethologist, the geneticist, the molecular biologist, etc., who brings natural objects under his control by acting on them and talking about them and who organises and restructures them to his purposes.

(2) "*Construction*": Gradually, a strict *division between the production and utilisation of knowledge* is established, resulting in a "scientification" of the objects thus constituted in practices. Subsuming under "method" the cumulative regulation and standardisation of both linguistic (conceptual and syntactical-theoretical) and nonlinguistic (concerning observational and experimental skills) procedures, the *substantial purposes* of human practices to do with our dealings with conspecifics, an-

imals and plants can be distinguished from *methodological aims*: The linguistic means for the development and communication of knowledge are to be improved in order to yield coherent, consistent and explicit terminologies. Such an aim can be justified because the construction is oriented towards the scientific, i. e., the *trans-subjective controllability and validity* of its results. Ideally, both the linguistic expressions and the non-linguistic experiential basis would become independent from the individual human agent and could, in principle, be duplicated by each subject ("trans-subjectively"). Unlike the dominant descriptive philosophies of science which have been mentioned above, here we do not give prominence to a form of language, but also consider a form of poietical action and its development, resulting in the art of observation, experimentation, and measurement. In scientific contexts, these crafts require normative regulations which make possible, e. g., the impersonal reproduction of experiments and, thereby, ensure transsubjectivity on the poietic level, too.

With evolutionary biology in view, which – in sharp contrast with chemistry and large portions of physics – describes natural history, the *construction of past states of affairs* has to be considered. By definition, these are inaccessible to direct observation and experiment by contemporary biologists. Again, there are pre- and extra-scientific practices which serve as a basis for such a construction in the sense of a "scientification" of the narration of natural history. The archaic hunter, gatherer, or farmer interprets traces of life and other kinds of evidence (footprints, the remains of an animal's prey, deserted animal dwellings, indications of drought, a rockfall, etc.) in terms of his knowledge of nature. Similarly, in science as well as in everyday life today we interpret present states of affairs with the help of causal knowledge as "evidence" for past events which brought them about. The questions which give evolutionary biology its object are modeled on the same structure: In the light of our causal knowledge, which evolutionary processes have resulted in the diversity of life as we know it? (Already from the archaic roots of this mode of questioning it is obvious that "evidence" is not "given" but only forms in the light of the causal knowledge which the inquirer brings to his object. For example, someone totally ignorant of avalanches will not see any reason to interpret certain features on the slopes of a mountain as traces of such events, however conspicuous they may be.)

Past natural events which have not been witnessed and reported by human beings cannot and should not be represented as if they were recalled by someone who had observed them from a fictitious Archimedean point, situated nowhere in particular and at the same time everywhere. Rather, even with the best modern sciences at hand, they remain hypothetical. Criteria for the validity of such hypotheses are only available if we take recourse to the transsubjective testability of statements concerning present states of affairs and of the retrodictions based on these and transsubjective causal knowledge.

(3) "*Reflection*": The constitution and construction of the objects of a scientific discipline are complex cultural accomplishments and, as any human action, potentially subject to failure. Therefore, they have to be *considered as human actions* are. Their actual purposes have to be extricated and their success or failure have to be evaluat-

ed in the light of these. For contemporary biology, this means that its procedures for the formation of concepts, models, and theories as well as those methods commonly comprehended as "experiential" such as observation, experiment, and measurement are to be represented with the help of a scheme of means-ends-rationality and tested with respect to it.

For biology, this means that only at the level of reflection concepts, principles, and rules are brought into play which cannot be determined predicatively, i. e., by reference to objects or events which serve as examples and counterexamples. Quite obviously, organisms, species, genomes, or genes are not the kind of objects which someone could literally point his fingers at in order to distinguish them from other objects. They can only be conceptually determined within the context of linguistic considerations which do not only take natural or laboratory objects into account, but also refer to the linguistic and non-linguistic procedures of biologists themselves. It is a common misunderstanding which receives support from the modern, linguistically oriented philosophy of science, that the *concepts and theories* of a science like biology *refer exclusively to its subject matter* (to organisms, their structures and functions). However, even fundamental concepts of biology cannot be explicated without reflection upon the activity of biological research. This can be demonstrated by many examples which are central to the concept formation of modern biology. In short, reflection as an indispensable step in the "scientification" of extra-scientific practices relates the constitution of biological objects to the methods of the scientific production of knowledge.

The three steps of constitution, construction, and reflection together comprise a proposal for an improved understanding of biology which has already been put to use in biophilosophical publications. By these its fruits it should be evaluated. (In this volume, this is exemplified by the contributions of M. Gutmann and C. Hertler.)

6. Philosophical and biological conclusion

The programme for a methodical reconstruction of biology which has been sketched above aims to reveal our *access to the objects of biology through human action*. It does not proceed from a philosopher's fiction of biology but rather from the biologists' biology. On the other hand, it does not presuppose its validity (nor, for that matter, its failure). Thereby, dogmatic pre-decisions (e. g., concerning the existence of natural objects or states of affairs) and ideologically motivated predilections (e. g., for or against specific empirical methods or kinds of theory) are avoided. This helps to steer free from the realistic, naturalistic, or biologistic assumptions repeatedly evident in the contemporary discussion which do not withstand epistemological criticisms. The objects and results of biology are at the biologists' disposal inasmuch as they are determined by human purposes and their choice of means. In this sense, philosophy of biology can hand back the science of biology to the biologists. Even more sharply, one might say that successful *biology does not approximate its natural object but its cultural purposes*.

A philosophical understanding of biology which is both conceptually clarified and discursively well-founded can help to purge the biologists' view of themselves of its amateurish, *ad hoc* components which are of dubious origin and carry countless consequences whose importance is easily underestimated. Instead of giving nothing more than an *ex post* pseudo-justification of actual scientific practices and the received opinion, it could contribute to the elucidation of unresolved intra-biological controversies.

To put the matter less presumptuously: The philosophical reconstruction of biology offers a tool kit for the explicit clarification of the concepts, methods and theories of biology. This helps to redraw the widespread picture according to which the latter result from a quasi-natural, cultural history during which they are produced by unintentional mutations and selected for or against by the forces unleashed by unendingly warring factions.

The status of organism:
Towards a constructivist theory of organism

MATHIAS GUTMANN

0. Introduction

It is almost impossible to find a term in biosciences more controversially discussed than the term "organism". Within a wide range of concepts usually presented, some typical positions can be pointed out, to determine the objections regarding the methodological and ontological status of that term:

1. If an "organism" is supposed to be the unit of biology, it is fair to assume the presence of ontological implications, usually designated as "essentialist". The *essence* describes the organizing center ("Wesen", ousia in terms of Aristotelian approaches) of organismic characters[1]. This concept can be developed logically in two directions:

a) The essence can be described as a *static* entity, as a simple ontical representation of logical relations (e.g. the whole/part-relation or the element/class-relation). Within the resulting range of these arguments e.g. Troll's approach of an *Idealistic Morphology* should be placed.

b) The essence may be interpreted as a *process*, developing or *unfolding* the preconcepted possibilities of its definition. This form of a "process-oriented" platonism is presented e.g. by Whitehead's concept (Whitehead 1984).

The ontological point of view can be extended, describing the "nature" of organism by stating analogies such as the "machine analogy" (Gould 1970, Gutmann 1995), their description as thermodynamically defined "open" or "closed systems" (e.g. Prigogine 1985), etc. In this case the difference between the model (e.g. the thermodynamic system or the mechanical construction) is only given in terms of ontology and not of methodology; i.e. a class of thermodynamic systems or biomechanic constructions is given and the organism is described as a sub-class of the respective class.

2. A nominalist approach determines the "organism" to be just a word[2] which does not refer to anything in nature. The organism becomes an object of the decision of the scientist, to designate a group of things (living beings, plants, animals etc.) arbitrarily with a common *name*. This "nominalist" approach[3] can be applied on each

---

1   For a comprehensive discussion of the problems of essentialism in evolutionary biology see e.g. Mayr (1984 & 1997).

2   Johannsen introduces the term "gene" as a nominalist construction, representing pure-line-breeding-groups of plants.

3   The difference between "nominalism" and "realism" seems to be a rather outfashioned and inadequate because it simply denies the development of theoretical philosophy during the last

level of "biological organization", defining tissues as "groups" (or sets or classes[4]) of cells, organs as "groups" of tissues, organisms as "groups" of organs. This approach – despite the fact of its nominalist nature – obviously refers to a *realistic* understanding of the "structure" of biological "organization" or hierarchy.

3. From a realist point of view and in very contrast to the nominalist approach, the concept of organism is supposed to represent a "real aspect" of the "world". Usually such ontological interpretations of scientific notions lead toward a "hierarchy" of distinct levels of biological organization. According to such an approach, the organism is described as one of those levels (Bunge & Mahner 1997, Ghiselin 1987, Eldrige 1985).

4. Finally, organisms can be identified *directly* with any living entities. They are usually defined by one or more important characters; their being "alive", their "reproductivity", "irritability", their "having structures" etc. are traditionally given examples.

In all these cases the possibility to describe the respective characters either realistically or nominalistically, should be mentioned. Accordingly the characterization of a theory *as* an essentialist or a nominalist approach needs not be *completely* disjunctive. The situation may become even more complicated, if one takes into account the fact, that the characterization of a theory as a nominalist or realist approach is not the only methodological option; e.g. the pragmatic approaches provide an alternative, which has been widely neglected until now, particularly in biology.

Irrespective of the fundamental differences between this four approaches, they refer to the *same* concept of experience. They describe scientific methods as a more or less "adequate" description of pre-existing entities – indifferently to their specification as "natural kinds". Disregarding these principle aspects of ontology, the status of organisms was controversial during the history of modern evolutionary thinking, often confusing "organism-centered" approaches with essentilialism. Even Darwin, usually supposed to be an "anti-essentialist" biologist (Mayr 1984 & 1997), lines out the importance of the organism for evolutionary theory (Darwin 1897. s. Gutmann 1996). Following "Neodarwinist" approaches, the organism seems necessarily to be replaced by describing reproductive units in terms of population genetics alone. Reducing the "phenotypes" of living beings to the "genotypes" constituting the "gene-pool" of a population, evolution is reduced to the shift of the genetic composition of populations under the impact of selective forces – might they be the result of internal or of external "causes".

The rejection of essentialist concepts of organism is tightly connected with the problem of development and transformation. Following Mayrs argument (Mayr

---

100 years, particularly considering the so called linguistic and the pragmatic turn. Nevertheless, the contradiction just mentioned seems to govern nearly the complete "species"- debate during the last 40 years. For a more adequate reconstruction of the different "species"-concepts s. Weingarten & Gutmann (1993), Gutmann (1996) and Gutmann & Janich (1998).

4   In fact, the logical difference between "groups" as "sorts" of things in a everyday-lifeworld - language, classes (constructed invariantly to relations of equivalence) and sets should be mentioned. Within the given context, it seems to be unnecessary to draw a systematically relevant distinction here.

1997) the essentialist definition of an organism *seems* to exclude the possibility of development (and a fortiori the possibility of evolution) because an essence is logically defined in terms of *unchangeable* characters. Obviously this conclusion cannot be drawn necessarily, because the term essentialism can be reconstructed as a metapradicate, which describes a form of an argument, that deals with a special concept of substance. Consequently, there is no *logical* connection between essences and "unchangeability" on the one hand, "organism" and "essence" on the other hand, and thus, even population-oriented concepts can be essentialist concepts. Especially the "New Synthesis" can be found to be an essentialist and at the same time a naturalist approach of species and population (Weingarten 1992, Gutmann & Weingarten 1994, Gutmann 1996)[5].

Some prominent modern approaches, namely the structuralist one, place the "organism" into the theoretical center of biology – in contrast to Neodarwinism. Following the structuralist approach (existing in a variety of partly differing concepts) the organism is supposed to be an essential condition for the identification of developmental regularities and principles. Thereby, they avoid the reduction of "development" to "embryology". Indeed, also evolutionary ideas are discussed, leading e.g. to an organism-centered version of a rational taxonomy (Goodwin 1997). This taxonomy of "developmental types" defines the "invariants" of evolution. In this case the term "structure" does not imply unchangeable characters but sets of "intrinsic" laws, governing development *formaliter*, not *materialiter*. The advance of such a structuralist approach is evident as it provides a scientific basis not only for the reconstruction of the constitution and changes of embryonic structures but also for the comparison of those formative processes, dealing with taxonomically *different* forms. Often, those organism-centered approaches tend to connect the "organismic" aspects with Anti-Darwinism (e.g. Webster & Goodwin 1982). On the other hand, structuralist approaches resemble formalist ideas, probably adequate for the solution of embryological but not for the solution of evolutionary problems (see Gutmann & Voss 1995).

However, and apart from the problem whether a theory of organism must necessarily be Anti-Darwinism, the status of important biological objects, usually dealt with in biological theories of evolution as well as of ontogenetical development remains unclear, as the great debate concerning the "species-" and the "population-problem" show. In all these cases, may there be offered nominalist or realist propositions for their solution, the underlying problem, the *origin* of scientific objects – and according to this of an *adequate* concept of experience – remains unsolved.

This paper aims at an introduction of a concept of organism, which can hold a central position within a "rational biology" according to *protobiological* principles[6].

---

5   The application of "population oriented concepts" does not imply anything considering the ontological status of these concepts. Populations can be understood in terms of "nominalism" as well as of "realism" or "essentialism" or "pragmatism" etc.

6   The term "protosciences" designates the corpus of those methodological principles, which are applied to constitute the objects of the respective science; accordingly protophysics allows the foundation of physics, protochemistry the foundation of chemistry etc. The protobiological approach will be explained later in this paper.

This foundation of biology by proposing a constructivist theory of organism will be exemplified with two case-studies. The first presents basic principles of a constructional morphology and the second some fundamental notions of a *rational genetic* (fully developed in greater detail in Gutmann 1996). Thus, the construction of a "rational biology" based upon a concept of organism could provide an alternative approach for biological and – a fortiori – evolutionary thinking. This rational biology avoids reductionism as well as naturalism or any form of essentialism.

1. The problem of experience

To realize the explanatory power of an alternative (which can be called a "culturalist" approach, to point out the main difference to "naturalist" approaches; s. Janich & Hartmann 1996), it is necessary to emphasize, that the four approaches mentioned above, can be reconstructed as "empiricist" approaches. According to Kambartel's reconstruction (1968) at least a few characters are shared by empiricist concepts, which have been referred to as the "two principles of empiricist reasoning". First, each form of knowledge is based on simple statements, describing existing (e.g. natural) objects. This is called the "empirical basis of knowledge". Second, it is assumed to be possible to construct a language, that can be completely reduced to those simple statements of the empirical foundation. This procedure could be called a "methodological reduction of knowledge".

   According to this point of view objects are given as natural entities (e.g. organisms, species, population as well as morphological structures, functions, ecological niches etc.), perceived or discovered by observation in nature or laboratory. For the empiricist point of view a reliable standard of the description is its adequacy with "nature". This concept will generate at least some methodological problems:

1. To answer the question whether an observation was adequate the criteria of adequacy must not be given in terms of observation again. Consequently, a *correspondence* between the observation itself and nature must be defined, which generates the same problem on the next level of explication because reproductivity and transsubjectivity have to be ensured.

2. The conditions under which an observation is made adequately, producing adequate or correct observations, must themselves be defined *adequately*[7]. E.g. referring to reproductivity could generate a vitious circle.

   The methodological shortcomings, which are shared by empiricist approaches, are ultimately a result of a concept, which bases experience on *passive* aspects of receptions. Analyzing the expression "to observe" one should reconstruct this verb at least as a four-termed "predicate of action", because usually "someone" *observes* "something", referring to explicable "aims" and applying adequate "means". The same is true for the expression "to describe". The criteria of adequacy of an observa-

---

7   This is the case particularly regarding the relation between the "natural conditions" versus the "artificial laboratory conditions". It can be shown, that the so called "natural conditions" are not independent from the ability to produce and reproduce at least some of the conditions under which an observation ("in nature") takes place (e.g. Janich 1997).

tion (e.g. in the case of astronomy or microbiology) as well as of a description can be defined referring to the applied tools. If any element of the equipment is used, the norms of function of these devices allow to distinguish effects of the devices themselves from "actual" observations. Additionally, since observation itself is an action, the results should be reported in reference to the necessary actions and their order (this point will be explained in greater detail below). Finally, in order to *make* an observation or description a standardized language is needed, according to the aims, means and respective action (and their order) which constitute the respective observation. This "operation-based" approach of experience will work undoubtedly, even in the case of other, more common forms of experience. Each single scientific experiment represents a *succession* of more or less strictly ordered actions. The same is often true in regard to everyday-lifeworld experience. Lifeworld simply means "referring to human action, everyday communication and practices, skills which are potentially available[8] for everybody". In all these cases, experience is the result of human action or at least tightly connected with it. Analyzing human action one can generally discern the aims of the actor, which might be reached or not, and the means which are applied to reach these aims. According to this description, experience should not be misunderstood as a sum of passively received facts nor be reduced to perceptions which "happen to someone". Taking into account the importance of the aims a well as the means used to observe or describe something, or – generally spoken – to "make" any form of experience, a possible alternative to the analyzed empiricist approach can be given. The main differences may become clear, when we reconstruct the origin of scientific objects in reference to human action and not in reference to any supposed properties of "nature" or "natural objects". In a way this culturalist approach – in its very principles – resembles some crucial aspects of "interventionist" approaches (e.g. Dingler, v. Wright etc.).

2. The culturalist alternative: Reconstruction, constitution and construction

In line with a culturalist approach, science is understood as a specific form of successful or at least success-oriented human action. In contrast to other forms of cultural practices (derived from the greek praxis) like plant-breeding, building houses or driving cars, the performance of sciences requires the definition of very strict criteria of success. These criteria are usually seen in transsubjectivity, reproductivity of results etc. (see Janich 1989–1993). The *objects*, the scientist deals with, must not be seen as given naturally; these objects are the result of the scientist's action, *successfully* applying scientific methods in order to reach the previously defined goals. To gain a scientific object in terms of action (here designated as the *constitution* of the respective object), not only the aims and the means but in addition, the *beginning* of the action must be determined. As many other operations, a successful constitution of scientific objects, the introduction of notions into scientific theories will often depend on an *order* of the single steps of this action, which must often be

8   See for an elaboration of the argument: Janich (1989 – 1993).

observed very strictly (s. Janich 1995)[9]. This "principle of pragmatic order" (ppo) can be illustrated referring to human action, that will fail, if a specified order of steps is not observed. From the ppo, the "principle of methodical order" (pmo) is derived, describing the formation of scientific theories in terms of action. The pmo is a *rule* to construct a wellformed line of argument by applying only those notions and terms, which were introduced already into the respective theory. The pmo can be applied as a norm to ensure the success of the reconstruction of theories too. A reconstruction is a complex operation that covers at least the following steps, which must be performed in strict order according to the pmo:

1. Analyzing existing scientific theories. Usually theories show some typical short-comings, like a gap in the line of an explanation, a petitio principii or more pro-foundly a vitious circle.

2. The construction of a "methodological beginning". Comparable to many other forms of human operations a successful reconstruction depends on the selection of an adequate beginning. In contrast to a merely historically defined "matter of fact starting point" of a theory, its "origin" sensu verbis which remains methodological-ly contingent, a "methodological beginning" has to be found within everyday-life-world-practice. Accordingly, the "constitution" of fundamental scientific objects starts within "lifeworld". The resulting terms are of course thought to be the *first* terms in the line of an argument; they need not to be the *first* terms in the history of a respective theory. The result of the application of the pmo is a methodological reconstruction of the "normative" and not of the "historical" or "matter of fact" genesis of the respective theory (s. Mittelstrass 1974).

3. The construction of terms. To introduce terms into biological theories, non-bio-logical sciences (extended by physics, chemistry and other non-biological sciences) have to be applied to establish prescriptions and instructions for the description of biotic entities. For example, idealized breeding groups are used to determine the "genotype – phenotype" concept. Knowledge borrowed from engineering can be applied to determine functional and structural descriptions for morphological pur-poses (e.g. the famous "machine analogy" or the "hydrostatic principle"). As a con-sequence, the status of the resulting term can be determined in accordance to the method of their respective construction, e.g. as predicates, abstractors, ideators etc. (see for greater Detail: Hartmann 1993). Especially for biological purposes the ap-plication of the "constructivist model-theory" is particularly useful (Weingarten 1992, Gutmann 1996).

4. The further construction of scientific theories towards "rational theories".

Ultimately, the concepts and notions can be used to describe biotic entities *as* objects of biological theories. The result are different, well-substantiated biological theories, which are restricted in regard of their respective claim of validity, e.g. in the case of "rational morphology" or "rational genetics" etc.

---

9   Of course this will not be true for *every* single action or operation; even in the field of natural
     sciences a lot of operations can be found to be independent from such an order.

3. The standard language of description

Following this constructivist approach the most important everyday-lifeworld practice, which has been proven to be particularly useful for protobiological purposes, is the practice of breeding (especially of animal-breeding, plant cultivation etc.). In respect to this practice it is possible to talk about singular or collective actions, movements or qualities of animals or plants (living beings, "Lebewesen") as well-known objects, which can be described, used, manipulated and varied according to the aims of the breeder (described by the intended characters to be improved during breeding) and without referring to a *biological* theory. Those qualities, including animal action or movement, e.g. the non-scientifically described behavior like the mode of motion, food-gathering etc., can include aspects of the "Gestalt" of an animal, as well as the propagation of characters. The descriptions of those collective or singular qualities referring to cultural practices build the methodological beginning for the reconstruction of biological theories and the objects they are referring to; this reconstruction allows the determination of their respective methodological structure and status.

The resulting abstract notions are restricted to an explicable context of the argument. Two languages will result, a standardized language – containing all the abstractors, ideators etc. to be introduced into the scientific language (S) – and the language, that will refer to the everyday language (L). We can give a table as an open list, which contains all the introduced notions on the one side, their respective references in everyday life language and practices on the other. This might be exemplified for protophysical terms (see Janich 1989a):

everyday life language	protophysics
surface	plane
edge	line
corner	point

According to the purposes of protobiology a similar table can be constructed which refers to the everydaylife practices as breeding or cultivating, the keeping and utilization of living entities on the one hand, the resulting notions on the other:

everyday life language	protobiology
part of an animal	structure
utilization of a part	function (ref. to def. structure)
Gestalt of an animal	construction
breeding group	population
propagation of a breeding group	reproduction

For the single steps of theory-building several specified everyday life practices, engineering and e.g. physical or technical knowledge are applied as *models* to constitute and *structuralize* biotic entities *as* objects for biological theories. The term "structuralize" is used here to emphasize the operational definition of structures. Consequently, animals or plants don't "have" structures but they are becoming "structuralized" by applying models (s. below). To introduce the term "organism" in accordance to the culturalist approach the application of these models is necessary. In the following, an outline of the constructional theory of model application is presented.

4. The constructivist theory of model application

The empiricist concept of scientific objects and theories results in two distinct worlds: a world of basic propositions and a world of theoretical concepts. In natural sciences models are often understood as objects, sometimes formal objects, which represent natural entities. In comparison to the reference they are representing, models are considered to be less complex. Accordingly, models are described in terms of iso-morphic (two termed-) relations, linking two sets of well defined objects (the theo-retical terms on the one hand, the concrete objects on the other) unequivocally to-gether. Consequently several, semantically different interpretations provide true statements according to one and the same theory.

This approach raises the problem that no principles for the *construction* of the models can be given; models must be understood as idealized representations of empirical data, natural entities etc., but the *way* to *construct* such models, the crite-rion for the proper scientifically sufficient *application* of those models cannot be given, neither in terms of empirical data nor in accordance to underlying theoretical concepts alone. To do so would necessarily lead to either an regressus ad infinitum or a vitiosus circle.

In very contrast to this model theory, a constructivist approach does not imply an analogy or a representational relationship between the intended[10] natural object or process to be described or explained, and the knowledge applied as a model. An analogy is usually constructed in the sense of a two-handed relationship of immedi-ate representation – expressed for example in the machine analogy of organismic structures, or the similarity between ideal gas and genes or genetic particles of in-heritance. Within the constructivist approach this analogy is replaced by a rule-guided comparison.

A model is applied in the sense of a comparison, i.e. a three-handed relationship is constructed, comparing two things, e.g. the mammalian limb with a technical lever construction. The tertium comparationis then describes the *rules* governing the function of an idealized lever construction. The model as a "model for" an in-tended object represents the prescriptions and instructions applied to construct or introduce the notions, needed to describe living entities "as" scientific objects.

5. Two case studies

Some principles of the culturalist constitution of scientific objects shall be exempli-fied, reconstructing two cases, which – traditionally – have nearly nothing in com-mon. Based on these two examples, a constructivist solution of the problem of or-ganism can be proposed.

---

10   I. e. the object that shall be modeled, the "modelandum".

5.1 The rational morphology

The results of the following reconstruction[11] of a "rational morphology" differ fundamentally from similar approaches like that of Cuvier and even from some modern approaches from a structuralist point of view (see, for example Ho in this volume, and Goodwin 1997 etc.).

5.1.1 The constitution of morphological objects

The methodical starting point can be identified within the numerous cultural practices, using e.g. animals for human goals to carry, pull or move loads. Traditionally, horses or cattle are bred to optimize such characters, which support the demanded tasks. To simplify matters the following example for the construction of primary morphological terms is described only for animals. Accordingly, these animals can be described *as being used* to produce force. In an abstract way, they can be described "as" force-generating units (Kraftaggregate). Consequently, their limbs can be described to provide optimal working conditions in order to use these forces most effectively. These limbs are described "as" force-transmitting and working-structure; e.g. the tendons will serve "as" force-transmitting, the fore- and hind-limbs "as" working-structures etc.

The next step is the construction of primary morphological terms. The description of animals will begin invariantly to a respective *movement* or *motion* of this animal. We call the *descriptions* of those motions (Regungen), like "swimming," "running" "digging" etc. of an animal, the *bionomal options* of the respective animal. The bionomy (a term in S) contains all description of motion and movement (in L) of living beings and must not be confused with their respective description e.g. "as" *locomotion* (a term in S). Bionomy has to be constructed as an open list of such descriptions to avoid the problems of (miss-)understanding the quantifiers "all" or "sum total" ontologically.

"Bionomal" descriptions of animals result e.g. in the description of animals "as" constructions in terms of biomechanics. The terms used to describe an animal "as" a construction are gained by applying e.g. the machine-model. As the parts of a machine will have to work together in terms of "force-closure" to ensure the mechanical continuum during working, the structures of a construction (the parts of an animal denoted "as" structures which refer to the description of the animal "as" a bionomal construction) will have to fit in terms of *coherence*. These operations, which describe an animal "as" a "force-generating-unit" and its parts "as" "force-transmitting" or "working-structures" accentuate a shift of language: from an ordinary, life-world-language (animal, working, part etc. in L) to a standardized – here: biomechanic – language (structure, function, force, transmission etc. in S). Consequently, coherence, as well as bionomy, don't describe characters of an *animal* in L but of a *construction* in S.

---

11   The term "reconstruction" describes the constructivist method to introduce a notion into a scientific theory; the three aspects of "analyzing given theories", "constituting scientific objects" and "constructing notions" are exemplified for the terms "gene" and "structure". For a more comprehensive elaboration see Gutmann 1996.

L	**S**
list of descriptions of movement	bionomy
connection	coherence
animal	construction(b)
utilization	function
movement/motion (e.g. swimming, running etc.)	locomotion

(b) = "structuralized referring to an explicable bionomal option".

The construction of these terms is done by applying the constructivist model approach mentioned above. Obviously, and in very contrast to an empiricist model theory, a constructivist approach doesn't imply an analogy or a representational relationship between the intended natural object or process to be described or explained, and the knowledge applied as model.

5.1.2 From part to structure

In contrast to an empirist point of view, within the constructivist approach the analogy is transformed to a three-termed "model-relation[12]", by applying non-biological – here physical or technical – knowledge to structuralize parts of an animal "as" morphological structures, e.g. the mammalian leg serving "as" a working structure is structuralizes according to a technical lever-construction. The tertium comparationis then is the *functional rule* governing the function of an idealized lever construction, for example in regard to its optimal geometry, the relation between the leverarm of a force, the lever-age of load, i.e. their relation to the supporting point. In comparison to a lever construction the mammalian leg can be *structuralized*, describing e.g. the bones as force-transmitting structures, which are functionally characterized by their flexural stiffness etc. In general, it is possible to structuralize the "Gestalt" of an animal under the given conditions as its respective bionomal construction, referring to the practice of preparation (anatomical sectioning) by applying the functional principles which are gained by modeling in the first step. The result of this model-based sectioning is the *functional description* of parts of the animal "as" the structures of the animal's construction. Applying functional models to structuralize a part of an animal, the resulting *structures$_F$* (the index F means "functionally") is the product of human operations using e.g. anatomical or histological methods. Of course, the "function" of a structure is not a "natural property" of this structure. We can call this operation, by which a part of an animal becomes functionally structuralized the "assignment of a function". When we refer to the different options of utilization, different "functions" of a structure and consequently differing structures might result – even, and this is an important point, if they refer to the *same part* of the animal.

In regard to the tables given above the results of the mechanically guided structuralizing of animals and their respective limbs or parts can be listed in a table:

---

12 Obviously considering its formal structure, a model-modelatum-relation is that of a constructional comparison.

(L)	**model (Mx)**	**result (S)**
part/limb	lever-arm construction	structure(Mx)
Gestalt	M1-Mn	construction(M1-n)

In addition, it is possible to describe even the mechanical interaction of several functionally defined structures referring to a particular mode of their action. If muscles are described as "tensible-force-generating structures" (Zugfaserelemente = TFGS), refering to the performance of the two TFGS on a single "force-transforming-structure" (i.e. bones described "as" FTS characterized here by their flexural stiffness), an *agonist* and an *antagonist* may be distinguished. In regard to the applied models several *types* of antagonism can be defined mechanically. These types of antagonism may differ relative to the geometrical arrangement, the mechanical properties of the constituting materials, their efficiency etc. Applying the hydraulic model other types of antagonism can be constructed. These antagonistic arrangements e.g. of TFGS working within a hydraulic (i.e. a fluid-filled) construction, thereby using the fluid itself as a "working-substance" to generate indirect antagonism, cannot be reduced to classical lever constructions, applied as models above.

5.1.3 From the Gestalt to the form

One of the most prominent problems of traditional morphology, namely whether the "form" may be an intelligible scientific object at all, can be solved, applying the method of rational morphology I have proposed here. After the description of animals in terms of structure, function, construction, coherence, bionomy etc. the term *form* can be introduced. Following this approach "form" describes a character of the construction of two animals, structuralized successfully invariantly to the same criteria of a mechanical description. If two animals are identical in reference to their construction, they share the same "form". The "identity of form" is an abstractor, introduced invariantly to the "identity of construction". Consequently, a methodological difference between taxonomy and systematic can be made. Systematic is just a specified form of a general taxonomy, by which in dependence of their construction animals are assigned to a particular class. Based upon such a systematic the reconstruction of evolution is possible (see Gutmann 1996).

In summary, all those terms, like structure, function etc. are gained by using e.g. engines or other machines as models to structuralize animals. Following this constructivist approach, scientific objects, constituting the "universe of discourse" of rational morphology aren't natural entities but *products* of human *operations*. Morphology itself will result as one of the numerous disciplines of biology, irreducible to any other discipline.

5.2 The basics of rational genetics

Traditionally, genetics is supposed to describe completely different phenomena than morphology. Nevertheless – and by no means only by authors of the "New Synthesis" – during the history of modern biology several attempts have been made to reduce morphology to genetics.

From a constructional point of view the scientific objects genetics deals with, are indeed completely different from those of rational morphology, but, they are as well formed and intelligible. The aim of this reconstruction is to provide a basis for a rational genetics. At the same time we will approach the question, whether there exist "biological laws of nature" comparable e.g. to the law of inertia in physics. According to the constructivist approach, even physical "laws of nature" can be reconstructed in terms of experimental operations (Janich 1997). A tight connection to the experimental conditions can be shown, to the "aim of knowledge" (Erkenntnisinteresse) which justifies the respective experiment, the applied devices (especially in case of measuring instruments) and the respective "laws of function". These "laws of function", *prescribing* the correct construction and application of the devices can be shown to be an important basis for the "derivation" of the "laws of nature". It is reasonable to realize an "*apriori* of the practice of measurement" for the constitution of physics (s. Janich 1989a & 1992). The proposed reconstruction of some aspects of classical genetics will show, that there are parallels to the situation in physics.

In contrast to morphology the constitution of genetics will begin with collective "properties" like *propagability*. Despite the fact that many plants can be propagated asexually with one single plant, the term "collective" seems to fit. Analyzing namely the term "propagation", at least determining a *successful* propagation, there must at least result more than one plant. To avoid misunderstandings propagation shall be used in the sense of "extended propagation"; preferring this meaning of propagation in terms of a collective property also the breeding of animals can be taken into consideration.

The methodological starting point of the constitution can be found again within the well known life-world practices of breeding animals or cultivating plants. To simplify matters and following some famous historical models, the constitution of a *rational genetics* is exemplified by the reconstruction of Mendelian cultivation. We base this reconstruction on Mendel's own report (Mendel 1901 & 1993), a reconstruction by Monaghan & Corcos (1990) and the constructivist reconstruction by Gutmann (1996). The parallels between the history of genetics and the methodological beginning is of course not a lucky coincidence between the "context of discovery" and the "context of justification". It is possible to assume that the form of cultivation, shown by Mendel's experiments, is indeed a "pragmatic *apriori*" of genetics – incidentally not only of classical genetics.

5.2.1 The first steps

To produce a Mendelian Population, first of all two pure-line breeding groups have to be established. They shall be characterized by well distinguished characters e.g. the color of the flower or the seed, the stem-length etc.. The modes of the propagation of these characters and the properties of the characters themselves have to be defined in terms of "aims of cultivation". Accordingly, the two fundamental modes of heredity, the alternative and the intermediate form, represent different explicable aims of cultivation.

Having established the intended breeding groups, cross-breeding will result in the hybrids of the generation f1. The uniformity of generation f1 can be expected, when idealizing the aims of breeding of the first step. Under these conditions it is possible to discern the *successful* from the *unsuccessful* cultivation. The same is true for the fission of the characters in the generation f2. Again, based on the aims of breeding, one can distinguish between successful and unsuccessful cultivation. The aims of breeding represent the "criteria of success" even *before* one single step of cultivation or crossing is done. Cultivating plants and breeding animals are undoubtedly forms of human action. Consequently, the single step of cultivation, crossing, pure-line-breeding etc. refer to *instructions* ensuring their respective success. This instructions can be applied to construct some fundamental terms of genetic theory.

5.2.2 From a "cultivation- or breeding-group" towards a population

In regard to this instructions that must be obeyed to gain a pure-line breeding group with the intended properties of characters, all those breeding groups produced in the same way in accordance to these prescriptions can be defined as having the same pre-disposition ("Anlage") or, to use an untranslatable German term, as being "anlagengleich" (identical in reference to their predisposition). The identity of disposition is a relation of equivalence (characterized by symmetry, reflexivity and transitivity). Arguing invariantly to a relation of equivalence, the abstractor (abstract term) "gene" in the sense of "genetic identity" can be introduced. Consequently the term gene refers to a standardized prescription of breeding group production, and does not belong to the same language as "disposition", which denotes the relation of equivalence. Each breeding group described invariantly to genetic identity is defined as an "ideal" population. We call this a Mendelian or I-population. For those populations a "norm of selection" (the word selection without any biological meaning here) can be derived, in dependence on the instructions of production: this norm is the Hardy -Weinberg law. The resulting standardized language describes scientific entities in reference to the explained practice of generating breeding groups described in L. Following the constructivist approach the I-population can itself serve as a model for the construction of terms of population-genetics (see Gutmann 1996).

Even if we choose the same methodological starting point, relating to the same cultural practice, the resulting biological theory deals with completely differing objects, because the "bionomal" properties of propagation *cannot be expressed reasonably* in terms of morphological description. The resulting biological theory is consequently not reducible to morphology and vice versa. Rational genetics, which I could describe here only very briefly also provides some powerful tools to reconstruct more complex theories dealing with objects in terms of molecular biology. The operational reconstruction of these objects requires a constructional concept of experiment which – until now – exists only in its very foundation.

6. The impact on the problem of the organism

The two case studies enable us to construct the term "organism" as a protobiological concept. In contrast to the organism-centered concepts mentioned above the notion "organism" can be introduced into theoretical biology as a "concept of notion" (Verstandesbegriff, Reflexionsbegriff) when this concept is applied in a similar way as in the case of "space" in protophysics (Janich 1989a). Accordingly, the "organism" is constructed beginning with single or collective properties or characters of living beings (animals, plants etc.). These properties or characters are then used to constitute primary biological terms, which we can describe as "organismic properties". In reference to these organismic properties we can use the term "organism" as an abbreviation, listing all those properties or characters of living beings described "as" organismic properties. According to this point of view, the term "organism" is part of a "meta-language" M, which is invariant to Sx, and describes the constituted primary and the constructed terms of biology, referring to everyday life practices in L. This "summary" of organismic properties should be seen as an open list of appredicates (the organismic properties), which can be completed if necessary (e.g. in the case of new biological disciplines).

In summary, when we can reconstruct the term "organism" in the suggested culturalist way, it gains some crucial characters, which sets it apart from other approaches:

The notion "organism" is substituted by "organismic properties". The "organismic properties" can be defined operationally constituting an open list of "appredicators", describing the bionomal options, which are gained by the model-based constitution of the "universe of discourse" of specified biological theories beginning with the qualities or movement of animals and plants. The methodological reconstruction of biology leads to distinct and well defined branches like functional morphology, constructional morphology, reconstructional morphology, constructional genetics, constructional theory of evolution etc. The "functional" and "constructional" description of animal determines an "organismic construction". The same is possible in regard to collective properties like propagability.

When regarding the organism as an concept of notion ("Reflexionsterminus"), the *fallacy of misplaced concreteness* can be avoided and in the same time the organism will regain its central place in biology as has often be seen as a necessary precondition for a rational biological science. Johannsen called the term "gene" just "a very applicable little word". In the same sense we can call the term "organism" a "little word". However, it is *applicable*!

7. Literature

Darwin, Ch. (1897): Origin of Species. Vol. I & II. New York.

Eldredge, N. (1985): Unfinished Synthesis. New York, Oxford.

Ghiselin, M. T. (1987): Species Concepts, Individuality, and Objectivity. Biology and Philosophy, **2**:127–143.

Goodwin, B. (1997): Der Leopard der seine Flecken verliert. München.

Gould, S.J. (1970): Evolutionary Paleontology and the Science of Form. Eart. Sci. Rev.; **6**: 77–119.

Gutmann, M. (1996): Die Evolutionstheorie und ihr Gegenstand. VWB, Berlin.

Gutmann, M. & Janich, P. (1998): Species as cultural kinds. Towards a Culturalist Theory of Rational Taxonomy. Theory in Biosciences, **117** (3): 237 – 288.

Gutmann, M. & Voss, T. (1995): The disappearence of Dawinism – oder: Kritische Aufhebung des Strukturalismus. In: Rheinberger, H.J. & Weingarten, M. (Hrsg.), Jahrbuch für Geschichte und Theorie der Biologie. **2**: 189–212.

Gutmann, M. & Weingarten, M. (1994): Veränderungen in der evolutionstheoretischen Diskussion: Die Aufhebung des Atomismus in der Genetik. Nat. u. Mus. **124** (6): 189–195.

Gutmann, M. & Weingarten, M. (1995): Die Struktur des systemtheoretischen Arguments. In: Rheinberger, H.J. & Weingarten, M. (Hrsg.), Jahrbuch für Geschichte und Theorie der Biologie. **2**: 7–15.

Gutmann, M. & Weingarten, M. (1996): Form als Reflexionsbegriff. In: Rheinberger, H.J. & Weingarten, M. (Hrsg.), Jahrbuch für Geschichte und Theorie der Biologie. **3**: 109–130.

Gutmann, W.F. (1995): Die Evolution hydraulischer Konstruktionen. Frankfurt a. M.

Hartmann, D. (1993a): Naturwissenschaftliche Theorien. Mannheim, Wien, Zürich.

Janich, P. (1989a): Euklids Erbe. München.

Janich, P. (1989b): Der Natur nach konstruieren. Erkenntnistheorie und Anwendung. In: Otto, F. et al. (Hrsg.): Natürliche Konstruktionen, Beiträge zum Internationalen Symposion des SFB 230 Bd. II, Stuttgart 1989: 47–55.

Janich, P. (1992): Grenzen der Naturwissenschaften. München.

Janich, P. (1993): Die methodische Konstruktion der Wirklichkeit durch die Wissenschaften. In: Lenk, H. & Poser, H. (Hrsg.), Neue Realitäten – Herausforderung der Philosophie: 460–476. Berlin.

Janich, P. (1995): Konstitution, Konstruktion und Reflexion. Zum Begriff der »methodischen Rekonstruktion« in der Wissenschaftstheorie. In: Demmerling, Ch., Gabriel, G. & Rentsch, T. (Hrsg.): Vernunft und Lebenspraxis: 32–51. Frankfurt a.M.

Janich, P. & Hartmann, D. (Hrsg.) (1996): Methodischer Kulturalismus. Frankfurt a. M.

Johannsen, W. (1911): The Genotype Conception of Heredity. Amer. Natur., **45**: 129–159.

Kambartel, F. (1968): Erfahrung und Struktur. Frankfurt a. M.

Mayr, E. (1984): Die Entwicklung der biologischen Gedankenwelt. Berlin, Heidelberg, New York, Tokyo.

Mayr, E. (1997): This is Biology. Cambridge.

Mendel, G. (1901): Versuche über Pflanzenhybriden. Leipzig.

Mendel, G. (1993): Experiments on plant Hybrids. Übers. von Sherwood, E. R., Corcos, A. F. & Monaghan, F. V. (eds.). New Brunswick, New Jersey.

Mittelstrass, J. (1974): Die Möglichkeit von Wissenschaft. Frankfurt a. M.

Prigogine, I. (1985): Vom Sein zum Werden. München, Zürich.

Webster, G. & Goodwin, B.C. (1982): The origin of species: a structuralist approach. J. Social. Biol. Struct., **5**: 15–47.

Weingarten, M. (1992): Organismuslehre und Evolutionstheorie. Hamburg.

Whitehead, A.N. (1984): Prozeß und Realität. Frankfurt a. M.

Explanations in a historical science

WALTER J. BOCK

Introduction

Analyses of explanations in biology have been hindered because different, although partly overlapping explanatory systems have been proposed and because biologists and philosophers of science have not been clear on what constitutes acceptable and/ or complete explanations. Such understanding is especially difficult for historical aspects of biological features. Although the primary goal of this paper is to inquire into the nature of explanations in historical biology, it is necessary to examine general systems of explanation in biology, to investigate the relationship between different systems of explanations in biology, and to discuss what constitutes full explanations in biology. In order to avoid any misunderstandings, two comments are needed. First is that I accept the existence of law-like statements in biological explanations, including evolutionary biology. Second I am not concerned as to whether biology is or is not autonomous from the physical sciences. I do not consider explanations in biology to differ fundamentally from those in the physical sciences. The only differences is that the physical sciences are rather simple compared to the biological sciences and that any philosophy of science based solely on the physical sciences is too simplistic to be realistic. The basic thesis to be advocated in this paper is that all evolutionary explanations are dependent on prior functional explanations – that is, functional explanations precede all evolutionary explanations, both nomological-deductive and historical-narrative.

Explanations

Explanations in biology deal with phenotypic attributes of organisms and their genetic basis, including relationships between phenotypic attributes found in the same or different organisms (e.g., species) as well as correlations between phenotypic attributes and diverse aspects of the external environment. Hence, given any phenotypic attribute, a complete explanation includes resolving: (a) all existing physical-chemical properties (form, function, etc.) which are functional explanations; (b) its ontogenetic development resulting from interactions between the existing genotype and the external environment (programmed systems, Mayr, 1974; 1988; 1997) which are also functional explanations; and (c) its evolutionary origin and, therewith, the evolutionary origin of the genotype, which constitutes evolutionary explanations (see Bock and von Wahlert, 1963 for an early discussion of diverse evolutionary explanations). Throughout, I will use functional explanations in the general sense of functional analyses in biology (Mayr, 1982), not in the sense of functional explanations in philosophy (see Nagel, 1961). The latter concept really explains nothing within the scope of science and should be dropped as part of scientific explanations.

The system of functional versus evolutionary explanations in biology is used as equivalents of functional biology versus evolutionary biology (Mayr, 1982:67–71 and of proximal versus ultimate causation (Mayr 1961). Unfortunately a blurring between these two areas of biology exists in Mayr's The Growth of Biological Thought which purports to deal with '"ultimate" [evolutionary] causation', p. vii, but includes a large amount of functional biology, especially in Part III on Variation and Its Inheritance. I prefer the expression "proximal versus ultimate explanations" which is clearer in meaning than the usually used "proximal versus ultimate causation" (Bock, 1988). Functional and evolutionary explanations are both required for full explanations in biology. Functional explanations cover "how" questions in biology – namely how known attributes of organisms work or operate and how these attributes develop ontogenetically. Evolutionary explanations are the "why" questions in biology – namely why the known attributes of organisms came into being originally and have modified (= evolved) over historical time (e.g., longer than one generation). Functional explanations are independent of evolutionary explanations, but the reverse is not true. Moreover, it is simply not valid to conclude that "Nothing makes sense in biology except in the light of evolution." (Dobzhansky, 1973). Functional explanations not only make eminent sense in the absence of any and all evolutionary explanations, but they constitute the large majority of explanations within biology, both pure and applied.

The dichotomy of functional versus evolutionary explanations in biology is not the only possible explanatory system. A second and perhaps more comprehensive scheme is nomological-deductive (N-D E) versus historical-narrative explanations (H-N E) (Bock, 1988, 1991; Szalay and Bock, 1991). Both N-D and H-N explanations are theoretical statements regardless of whether they are called theories, hypotheses, concepts, laws, etc, and both must be tested against objective empirical observations. These two types of explanations differ sharply from one another and must be analyzed separately.

Nomological-deductive explanations are the "standard" form of explanations in science and the subject of most inquiries into explanations by philosophers of science. N-D Es are in the following form. Given a set of facts (e.g, initial and boundary conditions) and a set of laws (be they causes, processes, or outcomes; see Bock, 1993), both of which form the explanatory sentence, or explanans, a particular conclusion, or explanandum, follows (Hempel & Oppenheim, 1948; Hempel, 1965:335–338). N-D Es answer the question: how has a particular phenomenon [explanandum] occurred? N-D Es apply to universals (non-limited sets of phenomena), do not depend on the past history of the objects or the phenomena being explained, and their premises (the nomological statements) are assumed to be always true. In saying that N-D Es apply to universals, these explanations are not temporally-spatially restricted within the proper region of the phenomena which for biology is the earth, and more specifically the "surface" (= the upper part of the crust) of the earth. If an explanandum resulting from the conjunction of the set of facts invoked (the initial

and boundary conditions) and the set of general laws disagrees with empirical observations, then the N-D E is falsified, and one must search for the reason for the
falsification. Falsification means that the explanandum does not agree with independent, objective, empirical observations. But falsification does not automatically
imply that the general laws used in the explanation are in error, although this is a
possibility. Possibly, the set of initial or boundary conditions used in the empirical
test is in error, or the empirical observations are in error. Examples of N-D Es include clarification of oceanic tides using gravity and of phyletic evolution evoking
natural selection (nonrandom, differential survival and reproduction of organisms).

Historical-narrative explanations provide an understanding of the existing attributes of a particular set of objects or phenomena at specified points in time; these
explanations depend on the past history of these objects and they must use pertinent
N-D Es. These objects explained by a H-N E are singulars, not universals, and have
definite spatial-temporal positions. H-N Es are stated on a nondeductive basis with
the hope of reaching the most reasonable and probable explanation for the objects
studied. Five aspects of H-N explanations must be stressed, the first being the most
important: (1) These explanations are historical, which means that earlier events
affect later ones. Consequently in any H-N E special care must be given formulating
the analysis within the correct chronological order of events and changes. (2) They
are given on a probability basis of being correct (Nagel, 1961: 26). This is necessary
because these explanations frequently involve a number of often conflicting N-D Es
employed in the explanation and because of the uncertainty over initial and boundary conditions involved in the explanation. (3) H-N Es must be based on pertinent
N-D Es, and these N-D Es, together with the pertinent empirical observations, form
part of the chain of arguments used in testing the H-N E. (4) H-N Es are not general
in that a successful explanation for one phenomenon (e.g., origin of homoiothermy
in mammals) need not hold for a similar phenomenon (e.g., avian homoiothermy).
(5) Because of their complexity, the possible confusion between conflicting explanations and the difficulty in identifying valid confirming and/or falsifying tests, H-
N Es must be stated clearly and unambiguously. Failure to do this may preclude
meaningful tests or appraisal of rival H-N Es. H-N Es in biology include the evolution, phylogeny and classifications of organisms or their attributes – that is, anything related to the history of life.

Both N-D E and H-N E are scientific under the criterion of demarcation for
scientific explanations advocated by Popper in that they are both available for testing by falsification against objective, empirical observations. But the two modes of
explanations differ in the many ways of how they are expressed, tested, and used to
test other theoretical statements.

Being theoretical scientific statements, H-N Es are available to tests by falsification, but such tests are often extremely difficult and inconclusive. Generally H-N
Es are not tested by falsification (in spite of numerous statements in the literature),
but usually by confirmation with the addition of more and more corroborating support. This procedure is closely akin, if not identical, to induction in the strict sense of
that concept. Objections cannot be raised to inductive testing of H-N Es because
they are theoretical statements about a finite number of objects in contrast to N-D Es

which cover universals or an unlimited number of objects. Testing of H-N Es depends on argument chains involving pertinent N-D Es and a large number of background assumptions (hypotheses; many being initial and boundary conditions), and finally tested against objective empirical observations. One should proceed to the empirical observations as directly as possible although the argument chain is often complex. The empirical observations and their roles as tests, whether falsifying or confirming, should be designated clearly. Phylogenetic systematists who claimed that phylogenies and classifications fall under the strict Popperian concept of deductive science and its testing simply do not understand the spectrum of explanations in science.

Being a scientific theory, a H-N E can never be proven, no matter how convinced one may be of its correctness. No serious scientist today doubts the origin of australopithecines (and their human derivatives) from an anthropoid ape; yet it is incorrect to claim that the origin of hominoids from these apes has been proven.

Evolutionary N-D explanations include the causal theory of evolution, or the set of causal mechanisms of evolution plus the initial and boundary conditions needed to resolve the two distinct evolutionary processes of phyletic evolution versus speciation. H-N evolutionary explanations cover descriptive and explanatory phylogenies, biological classifications which are derived from explanatory phylogenies, historical biogeography, etc.

Unfortunately the set of functional/evolutionary explanations are not equivalent in any simple way to N-D/H-N explanations. All functional explanations appear to be N-D and all biological H-N Es appear to be evolutionary explanations. However, evolutionary explanations are not exclusively H-N; causal evolutionary principles are clearly N-D (see Bock and von Wahlert, 1963). Possibly one could claim that the causal N-D evolutionary theories are properly functional explanations, but this would clash, and quite rightly so, with firmly established traditions of the perceived bounds of evolutionary explanations. Misunderstanding would arise in the future if evolutionary explanations were redefined to include just H-N E and thereby making these two sets of biological explanations equivalent. When debating evolutionary theories, few workers have distinguished between N-D and H-N evolutionary explanations, resulting in much confusion. Because most earlier discussions and most contributors to this symposium have dealt with H-N evolutionary explanations, I shall stress this class of evolutionary explanations.

Levels of organization and reductionism

All biological organisms display levels of organization (= hierarchies) which means that one must be concerned with (a) explaining biological properties at each and every level and (b) proper extrapolation of explanations from one level to another (Bock, 1991). Organization of smaller units into larger units is essential and must always be considered fully in any biological explanation. The exact organization of smaller units into larger ones determines the possible extrapolation of explanations from one level to another. Properties of units at any level of biological organization depend on the properties of the subunits and on the exact organization of subunits into larger units which explains the "emergence" of new properties in the larger units not existing in the smaller units – often expressed as "the whole is greater than the sum of the parts".

No basis exists for any assumptions that different levels of biological organization have different explanatory systems. No justification exists on which to claim that lower levels of biological organization, e.g., the molecular and cellular levels, can be explained fully with the laws of chemistry and physics, but that higher levels of biological organization (e.g., the organism) cannot be so explained. I maintain that the same aspects of explanation exist at all levels of biological organization, including the molecular level. For example, at the molecular level, why only left-handed amino acids can be used by known biological organisms is an evolutionary explanation and cannot be resolved using the laws of the physical sciences.

With levels of organization comes the question of reductionism in which the reduced theory is explained completely by the reducing theory; two types of reductionism exist, namely: (a) interdisciplinary (e.g., can biological explanations be reduced completely to explanations in the physical sciences); and (b) intradisciplinary (or atomism, e.g., can all biological explanations on higher levels, such as the organismic, be explained by molecular biology). Interdisciplinary reductionism can be further subdivided into strong (in which the reduced theory can be derived completely and uniquely from the reducing theory) and weak (in which this is not possible). Herein I follow closely Nagel's (1961) treatment of reductionism in which he has shown that strong reduction is so exceedingly rare in science that it can be ignored, and all interdisciplinary reductionism can be treated as the weak form.

The answer to the question whether all biological explanations can be reduced to explanations at the lowest level of organization, i.e., molecular biology, is clearly no. There are important biological principles at every level of biological organization which depend absolutely on the particular organization of units at that level and which do not exist at lower levels. The objects investigated in all scientific disciplines, including physics and chemistry, have levels of organization in which new properties appear at successively higher levels simply because of the details of organization. Hence, claims about the possibility of complete intra-discipline reduction in any science appear to be baseless, and it would be best to abandon the atomistic concept of reductionism (= intradisciplinary). However, this does not mean that it is not essential to analyze properties at all levels of organization (Kramer and Bock, 1989). This requirement holds in all sciences in which distinct levels of organization exist. This is especially true for living organisms for which levels of organization are intrinsic. And it is not correct to confine biological investigations to the level of the organism, as there are higher levels such as individual-individual interactions, individual-environmental interactions of many forms, interactions between members of a ecological community, etc. Hence, my use of reductionism will be restricted to interdisciplinary reductionism, and then only to the weak form. It must also be noted that even if one accepts a strong holistic position, or if one is primarily interested in biological organization at the level of the organism, the most (and probably only) successful research strategy is a non-holistic approach in which the parts of the organism (or of the ecological system, etc.) are first analyzed at successively lower levels of organization to the lowest levels possible, followed by a rebuilding of the units at successively higher levels of organization until the original starting point is reached. Unfortunately the history of biological research has very largely been a continuous one of reductionistic studies as newer techniques and equipment became available and without the necessary reconstruction of parts to the original levels of the organism and organism-ecological systems.

Interdisciplinary reduction (e.g., reducing biological explanation to those in the physical sciences) is important for biologists, but this is restricted largely, if not entirely, to functional explanations in biology. Yet not all functional biological explanations are susceptible to reductionism by the physical sciences. Some explanations, e.g. the Hardy-Weinberg equilibrium and concepts of ecological competition, are not reducible to physical-chemical explanations of the existing universe because they do not depend on the materials comprising living organisms. Most, if not all, evolutionary explanations, be they N-D or H-N, cannot be reduced to explanations in the physical sciences because these evolutionary biological explanations again do not depend on the materials comprising living organisms and hence on the physical-chemical explanations of known matter or because they are historical and thereby lie outside of the scope of physical-chemical explanations.

If one considers reductionism as a tool in scientific study – that is, as the necessary analytic phase of inquiry – then it is not only an important research strategy, but an essential one. But if one considers reductionism as a philosophical program in science, then it is a completely invalid approach for biology and should be abandoned. This does not mean that a holistic approach should be used exclusively because the reductionistic analytic phase is essential in any inquiry.

Priorities of Explanations

It is not possible to present even a brief sketch of the history of the interrelationships between functional and evolutionary explanations since Darwin introduced his idea of organic evolution by natural selection in 1859. Suffice it to say that from 1859 to the present, functional explanations were confused with teleological explanations, and most evolutionists from the time of Darwin believed (erroneously, I should add) that the introduction of Darwinian evolutionary theory eliminated all teleological explanations from biology. This belief was interpreted by almost all evolutionists to mean that functional explanations were unimportant or unessential for evolutionary explanations; hence a rather complete chasm developed between functional and evolutionary biology which lasted until today to the detriment of both biological subdisciplines.

The first step in the appraisal of any evolutionary explanation, both causal evolutionary mechanisms and reconstructions of the history of life, must be an inquiry into the underlying functional explanations. If these are lacking or are not sufficiently corroborated by empirical testing, then caution should be employed before further considering the evolutionary explanations. If functional explanations do not sufficiently support a particular evolutionary explanation, than the first task is to investigate further the supporting functional explanations and to corroborate them by empirical testing. Only then it is possible to access the validity of the evolutionary account. This is not to say that a particular evolutionary explanation, be it a N-D or a H-N E, currently insufficiently supported by functional explanations is wrong; only that it should be used cautiously. Nor does it imply that causal evolutionary explanations properly supported by functional explanation are automatically valid; additional independent testing of the evolutionary theory is required. I do not take the position that only a single evolutionary theory exists, and that all competing theories are invalid. My concern is one of careful skepticism, especially of any evolutionary explanation not clearly founded on the necessary N-D functional explanations.

Causal evolutionary theories, such as the Synthetic Theory, are N-D Es, and hence depend on prior functional explanations. Assertions in the Synthetic Theory of genetic or phenotypic variation available every generation in a population are dependent on functional explanations in genetics. Details on the working of selective demands of the external environment on individual organisms are based on functional explanations such as functional/ecological morphology and organism-environmental interactions. Synthetic evolutionary theory can be subdivided into two different processes, namely: (a) phyletic evolution resulting from the dual causes of (1) the set of mechanisms for the formation of genetically-based phenotypic variation every generation and (2) the action of selective demands arising from the external environment, and on a complex set of initial and boundary conditions including (i) existing genetic variation in present population, (ii) properties of biological materials produced by the genotypic program, and (iii) geographic and habitat location of the population; and (b) speciation which depends on the sets of causes and of initial and boundary conditions for phyletic evolution plus the additional condition of the necessary external isolating barrier separating the two populations for a sufficiently long period of time to permit the evolution of intrinsic isolating mechanisms for genetic isolation. Synthetic evolutionary theory includes a number of additional N-D Es such as multiple pathways of adaptations, paradaptations, preadaptation, accident versus design, etc. Although it is not possible to go into details in this short presentation, I will assert that the fundamental components of the Synthetic Theory are sufficiently supported by well-corroborated N-D functional explanations.

H-N evolutionary explanations, such as classifications and phylogenies, are dependent on both causal N-D evolutionary theories and functional N-D Es. Unfortunately most biologists, regardless of which causal N-D evolutionary theory they accept, fail to base their H-N evolutionary explanations on proper and sufficient functional explanations such as functional and ecological morphology (Bock, 1981, 1990, 1991, 1992, 1994), behavior and ecology. As mentioned above this failure stems largely from a long standing confusion between teleological and functional explanations and on the belief that Darwinian evolution eliminated all teleological (and hence functional) explanations from evolutionary biology. Not only is the essential foundation of functional explanations disregarded, but many evolutionary biologists claim that the N-D evolutionary explanations must not be used as the foundation for H-N evolutionary explanations such as classifications and phyloge- nies. These workers claim that patterns in classification or phylogeny must be ascer- tained first from which causal processes can be determined. Yet nothing is said on how one chooses between the many possible patterns generated without the guid- ance of an underlying N-D explanation. Processes are not generated from patterns, but from conjectures. Subsequently, patterns in nature are generated and explained from the causal processes. Newtonian ideas about gravity was not ascertained from the pattern of tides, but was conjectured by Newton who then explained the tides based on his concept of gravity. Unfortunately, this almost complete exclusion of functional explanations from H-N evolutionary explanations has strongly under- mined the veracity of the latter and has led to the broadly held belief among biolo- gists for over a century that studies of classifications and phylogenies are useless speculations.

A notable exception to ignoring functional explanations is found in the exten- sive analyses of phylogenetic reconstruction of many groups of animals by mem- bers of the Frankfurt Group under the leadership of Professor Wolfgang Gutmann in that their H-N evolutionary explanations are firmly based on N-D functional expla- nations. Their approach to H-N phylogenetic reconstruction should be examined and considered carefully by other workers in their classificatory and phylogenetic analyses as an outstanding model of the proper approach to H-N explanations in evolutionary biology.

Conclusions

Central to all scientific endeavors is the explanation of observed empirical phenome- na; hence the core of any scientific methodology is an analysis of appropriate sys- tems of explanation. Biology, having a large historical component, requires a set of historical-narrative explanations in additional to more usual nomological-deductive explanations for a full explanatory system. A second major facet of biological ex- planations is the contrast between functional explanations of existing phenotypical attributes of organisms versus evolutionary explanations of the origin of phenotypic attributes over evolutionary time. Given any particular phenotypic feature, it is pos- sible to provide an explanation of its existing properties using only N-D functional explanations. Yet it is simply not possible to explain the origin of these phenotypic attributes using only N-D Es; H-N Es must also be employed. The history of biology over the past two centuries, especially the endless debate between mechanization versus the many forms of vitalism, is replete with attempts by both sides to achieve a complete explanatory system using only N-D Es.

Four important points must be stressed. First, is that functional explanations are prior to evolutionary explanations, including causal N-D ones, in that the former explanations must be established before it is possible to formulate evolutionary ex- planations. Second, not all evolutionary theories, as usually understood, are H-N Es or, in simple terms, historical. An important segment of these theories, including mechanisms of genetics, adaptation, speciation, etc., is strictly N-D. Third, it is simply not correct, as believed by many evolutionary biologists, that nothing makes sense in biology in the absence of evolution. Major portions of explanatory systems in biology (= functional explanations) are possible in the complete absence of any theory of evolution, Darwinian or otherwise. Indeed many to most biological explanations are proposed and used successfully in the complete absence of any evolutionary explanations. What would be lacking in the absence of evolution explanations is understanding the origin of given biological attributes and a full explanation of these biological features. Fourth, is that H-D Es must be founded on well tested N-D Es with the final testing being against objective empirical observations. It is simply not sufficient to use vaguely formulated N-D Es as the basis of H-N Es; rather one must test thoroughly the supporting N-D Es before formulating H-N Es in any science. Hence any H-N evolutionary explanation is absolutely dependent on well-tested and corroborated N-D Es, both functional and evolutionary.

Darwinian synthetic evolutionary theory, a N-D E, is broadly supported by well-tested functional explanations, such as the mechanisms of genetics, interactions between organisms and the external environment associated with the concept of adaptation, dispersal abilities of organisms over external barriers, etc, which does not claim that there are no aspects of current Darwinian causal evolutionary theory still lacking the necessary functional explanatory foundation. Darwinian synthetic evolutionary theory provides a necessary, but not sufficient, N-D foundation for all H-D evolutionary explanations. In addition, functional explanations are a second necessary, but also not sufficient, condition for all evolutionary explanations which cannot violate any well-corroborated functional explanations. The major conclusion to be offered is that if H-N evolutionary explanations are to have any scientific merit, they must be founded solidly on well-corroborated N-D functional explanations.

Acknowledgements: Firstly, I wish to thank Professor Wolfgang Gutmann for inviting me to this conference, but even more for the many years we had together discussing many interesting topics within and without biology and for the many walks we had along the south bank of the Main River. Further, I wish to express my appreciation and thanks to Professor Dominique G. Homberger with whom I have discussed this paper and who kindly read and commented on the manuscript.

Literature

Bock, W.J. (1981): Functional-adaptive analysis in evolutionary classification. Amer. Zool., **21**: 5–20.

Bock, W.J. (1988): The nature of explanations in morphology. Amer. Zool., **28**: 205–215.

Bock, W.J. (1989): Organisms as functional machines: A connectivity explanation. Amer Zool., **29**: 1119–1132.

Bock, W.J. (1990): From Biologische Anatomie to Ecomorphology. Proc. 3rd International Congr. Vertebrate Morphology, Netherl. Jour. Zool., **40**: 254–277.

Bock, W.J. (1991): Explanations in Konstruktionsmorphologie and evolutionary morphology. Pp. 9–29. In: N. Schmidt-Kitter & K. Vogel, ed. Constructional morphology and evolution. Springer-Verlag, Heidelberg.

Organism and morphology:
Methodological differences between functional and constructional morphology

CHRISTINE HERTLER

Following the constructivist approach the organism holds a central position within all branches of morphology (W. F. Gutmann 1989, W. F. Gutmann & Bonik 1981, Janich 1992, 1997, Weingarten 1992, 1993). The organism is considered as a reflexive and common point of reference, an open list to which different morphological specifications may be added (M. Gutmann 1996). It is therefore necessary to ensure that morphological statements refer to organisms and not to mere parts or aspects of them. This is accomplished by a methodologically fixed way of model application (M. Gutmann 1995): Models provide the required notions for morphological descriptions and explanations of living creatures.

However, the constructivist approach is not the usual context in which morphological studies are set. Usually, the main task of morphology is seen in the description and explanation of form and function of living creatures, and this task is differing methodologically from the reconstruction of organisms. This statement already implies that form and function belong to different morphological categories. Thus, both the different categories should be subject to separate study. Subsequently, for the purpose of describing the whole organism the results of these separate analyses have to be assembled. Partitioning the task of morphology in this analytical way has severe methodological consequences for the different morphological branches, the range of validity of their statements, and the way models are applied. Considering a specific case of model application these consequences become immediately evident. Therefore, the application of one important model, the hydrostatic skeleton model, is taken as an example and reconstructed in the context of different morphologies.

1. An analytical exclusion

The hydrostatic skeleton model is based on the assumption that fluids can serve as force transmitting parts as well as hard skeletal elements because of their incompressibility. In this way its application allows to describe the functioning of organisms or parts of them. Therefore, it can only be used in morphological studies which are indeed covering functional features of organisms. However, there are branches of traditional morphology explicitly excluding questions of functioning from their studies. This approach is widespread in taxonomy and systematics, which as a rule configure animals as assemblies of different characters. If such statements form the basis of a phylogenetic comparison, which is the topic of systematics in a traditional sense, two opposite reasons for the similarity of characters can be stated: Either the characters are similar because of their homologies (i. e. they point to a common ancestor) or they show similarity because of convergent evolution. The suspicion of convergence cannot be excluded where specific characters can be demonstrated to serve indispensable functions. For the purpose of a phylogenetic analysis it has to be ensured that only homologous characters, i. e. characters being essentially independent from function, are used.

According to the analytical partition of morphology in studies of form on the one hand and of function(s) on the other, the traditional method in taxonomy and systematics falls short of the second of the two mentioned aspects. It results in an analytically reduced description and therefore cannot supply with a sufficient basis for the reconstruction of evolutionary transformations.[1] Moreover, because questions of functioning are explicitly excluded from the morphological analysis, these kinds of traditional morphology do not allow the application of the hydrostatic skeleton model in configuring living creatures.

In other branches of morphology focussing on functional features of living creatures the hydrostatic skeleton model has proven its utility and fruitfulness (see e. g. Chapman 1975, Clark 1967, Elder & Trueman 1980, W. F. Gutmann 1972 and others). Two different forms of application shall be discussed here in detail and will be distinguished as functional and constructional approach, respectively.

2. An early career in functional morphology

The hydrostatic skeleton model was proposed during the 1950ies by Garth Chapman as a basis for functional descriptions of animal movement. He defined the term 'hydrostatic skeleton' as follows: "A hydrostatic skeleton may be defined as a fluid mechanism which in one way or another provides a means by which contractile elements can be antagonized." (Chapman 1958:338).

Essentially new in this definition was the proposed extension in the meaning of the term skeleton. 'Skeleton' had been used before mainly restricted to structures showing specific material characteristics: A skeleton consisted of hard parts, the exoskeleton of insects, for instance, or bones as parts of the vertebrate skeleton. 'Skeleton' is here defined materially. Chapman redefined the term 'skeleton' in a functional way as a means of muscular antagonism.[2] Although the hydrostatic skel-

---

1   These remarks on some methodological problems of character-based kinds of morphology should not be (mis-)understood as an adequate and complete methodological reconstruction; for further discussion see e. g. M. Gutmann 1996, Weingarten 1993 and others. As one purpose of this paper is to show how the hydrostatic skeleton model is applied in some branches of morphology it seems appropriate to mention morphologies, which do not rely on the hydrostatic skeleton model and to discuss possible reasons why it is rejected.

2   Chapman uses sometimes the expression 'muscular antagonism' and in other cases he speaks of 'contractile elements'. The term 'muscular antagonism' obviously restricts Chapmans version of the hydrostatic skeleton model to a zoological context and excludes its anticipation in botany. However in some cases 'muscle' is replaced by 'contractile element' to include cellular

eton model was then applied to reconstruct processes being already described, intestinal peristalsis during digestion for instance, the redescriptions on the basis of the hydrostatic skeleton model provided a new understanding of these processes.

With the aid of these definitions Chapman and his followers reconstructed lots of different movements of organisms, e. g. peristaltic burrowing of polychaets (Chapman 1950), several creeping modes of annelids (Chapman 1950, Elder 1980), jet propulsion of jelly fishes and cephalopods (Trueman 1980) and so on. After their reconstruction on the basis of the hydrostatic skeleton model, biomechanical questions were added and further experiments developed. For instance, Chapman and others tried to measure the changes of internal pressure within the coelomic chambers of annelids during different phases of movement (Chapman 1950, Chapman & Newell 1947). In the first instance Chapman searched for relations between a specific kind of hydrostatic skeleton, e. g. a *Lumbricus* or *Arenicola* type,[3] and corresponding modes of movement. Thus, hydrostatic skeletons of the *Arenicola* type suited very well for fast burrowing into smooth ground, whereas they are very slow in creeping. In the case of hydrostatic skeletons of the *Lumbricus* type it was just the other way round: Compared with *Arenicola* they were faster creeping than burrowing.

But within these interesting studies first problems arise with Chapman's specific functional application of the hydrostatic skeleton model. If *Arenicola* and *Lumbricus* are compared, differences in their realisation of movement can be recognized. But where do these differences come from? What is their functional cause? It can be noticed that the structure of the hydrostatic skeleton compartments is quite uniform in the *Lumbricus* type. Unlike that, hydrostatic skeletons of *Arenicola* type possess one large anterior segment, which facilitates burrowing because extensive changes of form are possible there and lead to a better anchorage within the ground. On the other hand, sideward bending of this large segment being necessary for creeping is impeded.

But this was not the conclusion drawn by Chapman. He addressed the differences in movement capabilities mainly to the specific kind of neuronal movement controlling equipment (Chapman 1958:346f). This peculiar discontinuity of the argument is resulting from the fact that Chapman understood his version of the hydrostatic skeleton model exclusively in a functional sense. Thus, it was only applied to reconstruct functional relations: Functional differences in movement capabilities were the point in question. Speaking about the relation of uniformity of segmentation and realisation of movement would have included a structural interpretation of the hydrostatic skeleton model, which has no explanatory meaning within a functional context. Thus, this argumentation demonstrates that the hydrostatic skeleton model was not used as a model for structural relations in a functional context due to the functionally restricted starting point of the study and the corresponding model application. This is a methodological explanation for the peculiar discontinuity of Chapman's argument, that is evident in other cases as well.

---

hydrostatic skeletons in particular. Thus, the question, whether the hydrostatic skeleton model in the sense of Chapman was formulated before in botany or not cannot be decided here.

3   From a systematic viewpoint *Lumbricus* is an oligochaete, and *Arenicola* a polychaete worm.

Returning to the analytical partition of morphology introduced at the onset, we are now able to recognize the functional reduction in this application of the hydrostatic skeleton model. In a character-based morphological analysis, living creatures are reduced to assemblies of characters having no functional meaning. Correspondingly, Chapman used the hydrostatic skeleton model to describe movement, a function, and thereby consequently excluded structural relations.

3. Functional types in systematics

Thus, the question arises how organisms are conceived within this kind of functional morphology. Considering the analytical method leading to the description of 'pure' functions it should be possible to assemble several functions – and in a wider scope the respective characters, under the condition that the relations between character and function are known. Characters and functions as they are found in nature[4] appear only in certain combinations. These empirically chosen assemblies are conceived as a set of indispensable organismal functions and serve as an example, a reconstructive norm for the organism. There are several problems arising with this functional concept of the organism: the empricial origin of the set of functions excludes an operational definition of the term ,organism' independent from empiric observation. Consequently, the completeness of the list of functions sufficient for an organism – its empiric definition – cannot be tested.

In 1975 Chapman published a list of several types of hydrostatic skeletons. He differentiated hydrostatic skeletons by two criteria: First by the structure – he spoke of open, closed and external hydrostatic skeletons – and secondly by the arrangement of muscle fibers in particular or contractile elements in general. Closed hydrostatic skeletons have no opening to the exterior and allow peristaltic as well as sinusoidal movements (e. g. segments of annelids). Body fluids and/or deformable tissues serve as force transmitting parts of the hydrostatic skeleton. Open hydrostatic skeletons support jet propulsion and have one or more openings to the exterior. Force transmission is carried out by the external medium; sea water in jelly fishes, for instance, replaces the function of the coelomic fluid in worms. External hydrostatic skeletons constitute only a minor subgroup and are not systematically explained. These groups are again subdivided in several classes by the arrangement of muscle fibers. Chapman distinguishes two forms of simple muscular geometry (one or two directions of contractile elements) and a complex muscular geometry (more than two directions).

Chapman noticed that his list, if it is used as a classification table in systematic sense would provide a classification that deviates considerably from usually used phylogenetic trees. For instance, the class of closed hydrostatic skeletons with sim-

---

4   This formulation includes a profound naturalism, which is doubted by the constructivist approach to morphological methods. From a constructivist viewpoint, characters and functions are both methodological constructs and, of course, cannot be considered as "found in nature" (e. g. P. Janich 1992, 1993, 1997, M. Gutmann 1996, M. Weingarten 1992, 1993).

ple muscle arrangement whose members are able to swim sinusoidally and/or to burrow enclosed several types of worms as well as chordates; the latter because the notochord itself represents a closed hydrostatic skeleton. Chapman noticed the empirical origin of his classificiation and the limited range of the differentiations. Methodologically spoken, this classification involved several functional types based on movement as far as the classes refer to several movement options. Because of the empirical origin it was neither evident whether the list was complete (whether all movement options were enclosed) nor whether these distinctions are related to structural differences in the hydrostatic skeleton model as a rule or accidentially.

4. Two questions in the context of evolution

Chapman did not use his hydrostatic skeleton classification to draw inferences on evolution. Considering the functional basis of these classifications, evolutionary inferences are impossible because analogous and homologous development cannot be distinguished without structural references. From a methodological viewpoint the construct 'function' represents the corresponding opposite analytical reduction as it has been already discussed (see part 1): If organisms are partitioned analytically in structural elements and functional relations, and both categories studied separately, evolutionary inferences on the development of organisms can neither be based on characters alone nor on single functions.

However, others (e. g. Trueman 1968, 1980) used similar classification systems for evolutionary purposes and tried to describe the evolution of burrowing or the evolution of jet propulsion. Those systematic and evolutionary reconstructions are of a very limited value for two reasons: At first, those descriptions are strictly limited to movements. Without any doubt there are other indispensable performances, like reproduction, feeding and others, which are completely excluded from the reconstruction. This is a problem of the functional starting point of the study. Secondly, those reconstructions are strictly limited to functions. As far as the functional approach results in functional descriptions of organisms, the question remains, whether such descriptions are sufficient to reconstruct evolutionary processes. This is a problem of adequate model application.

To find out what is meant by the term organism, we need to analyse how animals are treated as scientific objects in these functional studies. An organism is thus characterized by a specific kind of movement it performs with the aid of its hydrostatic skeleton. This is the difference generating tool which allows to distinguish between annelids and jelly fishes in the studies described above. The hydrostatic skeleton model is used as an instruction to structurize the description of movements. Consequently, an organism might be configured as a specific set of functions, functional types. Corresponding to the problems mentioned above, two questions remain but now with emphasis on the organism:

(1) How can we be sure that the list of functions enclosing movement, feeding, reproduction and alike is complete or not? How can we be sure that our description of the organism is complete?

(2) If we reconstruct evolutionary processes on the basis of our morphological de-scriptions, what methodological role do those descriptions play? How can we be sure that our descriptions are adequate for an evolutionary reconstruction?

5. Constructive extensions

At the end of the 1960s the hydrostatic skeleton model was applied by W. F. Gut-mann and adapted to the 'Frankfurt School' of morphology (1968, 1972, 1973). In this context its restriction to the description of movements was considered too nar-row and again it was reformulated. Presuming functional applications of the hydro-static skeleton model, Gutmann wrote in 1972: "In einem Hydroskelett herrschen nun leicht überschaubare Verhältnisse. Wenn eine gitterartige Muskelkonstruktion oder ein Schlauch vorliegt, so gehört alles was innerhalb und zwischen Muskeln vorhanden ist, als Füllung zum System. Das gilt auch für innere Organe, für den Darm."[5] (Gutmann 1972: 63).

This formulation completed and extended the application range of the hydro-static skeleton model in several aspects. Its application in a functional context is not excluded but the important relations (expressed in the definition mentioned above) are not functional ones. "Everything that is situated in and between muscles" is rather an expression for topological and thus structural relations in animal bodies. Obviously, this will lead to a different description, when animals are structured along the hydrostatic skeleton model. In a functional context, the description of annelid movement is finished, when two functionally different parts can be distinguished: a muscle layer (1) which acts on the coelomic fluid (2). In a constructively extended application this description is incomplete, because all tissues in and between the coelomic chambers are ignored, e. g. the intestine, vessels, neuronal fibers etc. This is the reason, why the term 'filling' is used in this definition instead of 'coelomic fluid'.

Describing the movement of animals in a constructive way allows to distinguish between three constructively different kinds of structures: force generating, force transmitting and working structures (M. Gutmann 1995). In relation to the annelid example, force generating structures are represented by all kinds of contractile ele-ments. The filling serves as a force transmitting structure. The whole set of struc-tures (the complete annelid body) is seen as a working structure. It is thus possible to describe movement as a coherent process. This kind of description is called recon-struction of locomotion instead of movement in a functional sense. Moving does not necessarily refer to an organism. If an animal 'moves its legs', the movement of the legs can be regarded as their respective function. Moving the legs is necessary but not sufficient to explain locomotion. For locomotive purposes legs have to be moved

---

5   Translation: The limitations and possibilities of a hydrostatic skeleton are easy to recognize. If
    a latticed muscle construction or a hose-like structure exists, everything, that is situated in and
    between muscles, belongs to the filling of the system. This is also true for internal organs, e. g.
    the intestine.

in a certain way[6]. Moreover, other parts of the body are also necessary as fixed points, against which forces can be transmitted to the substrate or medium in which the act of locomotion takes place. Thus, the explanation of leg movement is necessary but not sufficient for an explanation of locomotion. As an indication for that difference we will use the term *'movement'* in a functional context, while speaking of *'locomotion'* to express the reference to site changes of organisms in a constructive context. In general, this difference will be stressed by distinguishing between *functions* of parts and *performances* of complete organisms.

Moreover reformulating the hydrostatic skeleton model according to structures, i. e. to distinguish between force generating, force transmitting and working structures, implies that its application is not only restricted to the reconstruction of locomotion but also for other performances, like nutrition or reproduction. Now we are able to describe different ways in which organisms apply their hydrostatic skeletons. Consequently, annelids can be described as a framework of integrated and combined hydrostatic skeletons, one of which performs locomotion and the control of body shape and another one performing nutrition. The interaction of these different hydrostatic skeleton units (which evidently cannot be excluded on a mechanical basis) has to be configured in a way, that does ensure the functioning of all parts and units.

Those descriptions represent one type of construction. Another one can be demonstrated for jelly fishes, where the hydrostatic skeleton units for locomotion and nutrition are structurally identical but multifunctional. Constructive descriptions of animals have to be made in such a way, that the different hydrostatic skeleton units are demonstrated to follow the principles of coherence and bionomy[7]. This procedure ensures that they represent descriptions of organisms and not of single functions. A constructive reconstruction is completed, when it can be demonstrated that it follows both these criteria. From a methodological point of view these can be applied for testing the reconstruction.

6. Evolution and morphological reconstruction

On this constructive basis we are now able to reconstruct evoutionary processes. Due to the functional starting point and the analytical partitioning in functional studies, the problem occured to ensure the sufficiency of the functional descriptions for evolutionary inferences. It is possible to avoid this problem, if can be demonstrated that all descriptions refer to the organism. Both, the analytical reduction to function and the specific functional reduction to one mere function, can be solved methodologically.

The methodical starting points for evolutionary inferences are the morphological descriptions of organisms, which we call constructions. Since organisms have

6   A cat is able to move its legs while dreaming. Obviously, the cat is moving, but does not perform locomotion.
7   The constructive principles of bionomy and coherence will not be explained in detail here (see e. g. M. Gutmann 1995, 1996 and Hertler 1998).

actually changed morphologically during their evolutionary development, the rules
of these transformations have to be pointed out, which are the principles of optimization and differentiation. Two kinds of transformations are conceivable: either one
of the performances of the organism is optimized (principle of optimization) or the
range of use of one constructive unit is enlarged (principle of differentiation). The
reconstructive principles play a similar methodological role for the reconstruction
of evolutionary transformations as the constructive principles for the constructions
do: They allow to test transformation series and to distinguish between possible and
impossible series. These distinctions were introduced methodologically by M. Gutmann (1996) on the basis of adequate models, one of which is the hydrostatic skeleton model. We now may vary the constructions which represent the results of our
morphological studies according to those transformation rules in order to establish
preceeding constructions.

7. Summary

In the following the differences between functional and constructive approaches in
morphology will be summarized with reference to the organism.

   Within the functional approach the application of the hydrostatic skeleton model leads to descriptions of single functions; and therefore, organisms have to be
discussed as additive sets of functions namely functional types. Those lists cannot
be completed for methodological reasons, because a criterion for completeness is
missing.

   In the constructive approach the application of the hydrostatic skeleton model
leads to descriptions of organisms as constructions, which can be demonstrated to
work. This criterion for completeness on the level of constructions ensures that the
description refers to an organism.

   If we reconstruct evolutionary transformations on the basis of a functional approach we are restricted to single functions. This cannot lead to the reconstruction of
transformations of whole organisms because structural dependencies between functionally different parts are not included in these descriptions. In fact, as I pointed out
in the beginning this is due to the analytical shortcomings of this approach with
respect to the question of form.

   In the constructive approach these analytical shortcomings can be avoided, because we may constitute organisms and evolutionary transformations with the aid of
adequate models providing the required notions. This is exclusively true for a constructive, model-based approach.

Literature

Chapman, G. (1950): On the movement of worms. J. Exp. Biol. **27**: 29–39.
Chapman, G. (1958): The hydrostatic skeleton in the invertebrates. Biol. Rev. Soc. Cambridge **33**
       (3): 338–371.
Chapman, G. (1975): Versatility of hydraulic systems. J. Exp. Zool. **194**: 249–270.

Chapman, G. & Newell, G. E. (1947): The role of the body fluid in relation to movment in soft-bodied invertebrates: I. The burrowing of Arenicola. Proc. Royal Soc. London B **134**: 431–455.

Clark, R. B. (1967): Dynamics and metazoan evolution. – (Clarendon) Oxford.

Clark, R. B. (1980): The Nature and Origin of Metameric Segmentation. Zool. Jb. Anat. **103**: 169–195.

Elder, H. Y. (1980): Peristaltic mechanisms. In: Elder, H. Y. & Trueman, E. R. (Eds.)(1980): Aspects of Animal Movement. (Cambridge University Press) Cambridge: 71–92.

Gutmann, M. (1995): Modelle als Mittel wissenschaftlicher Begriffsbildung: Systematische Vorschläge zum Verständnis von Struktur und Funktion. Aufs. & Red. Senckenb. Naturforsch. Ges. **43**: 15–38.

Gutmann, M. (1996): Die Evolutionstheorie und ihr Gegenstand. Beitrag der Methodischen Philosophie zu einer konstruktiven Theorie der Evolution. (VWB) Berlin.

Gutmann, W. F. (1968): Die Evolution der Leibeshöhle im Wirbeltier-Stamm. Aufs. & Red. Senckenb. Naturforsch. Ges. **16**: 1–50.

Gutmann, W. F. (1972): Die Hydroskelett-Theorie. Aufs. & Red. Senckenb. Naturforsch. Ges. **21**: 1–91.

Gutmann, W. F. (1973): Versuch einer Widerlegung der Oligomerie-(Trimerie-)Theorie. In: Schäfer, W. (Hrsg.): Das Archicoelomaten-Problem. Aufs. & Red. Senckenb. Naturforsch. Ges. **22**: 51–101.

Gutmann, W. F. (1989): Die Evolution hydraulischer Konstruktionen. (Kramer) Frankfurt.

Gutmann, W. F. & Bonik, K. (1981): Kritische Evolutionstheorie – Ein Beitrag zur Überwindung altdarwinistischer Dogmen. (Gerstenberg) Hildesheim.

Hertler, C. (1998): Typus und Entwicklung – Varianten und Invarianten in der Evolution. In: Bucher, A.J. & Peters, D.S. (Eds.): Evolution im Diskurs. Eichstätter Studien (N.F.) XXXIX (Pustet) Regensburg:47 – 60.

Janich, P. (1992): Grenzen der Naturwissenschaften. (Beck) München.

Janich, P. (1993): Der Vergleich als Methode in den Naturwissenschaften Aufs. & Red. Senckenb. Naturforsch. Ges. **40**: 13–28.

Janich, P. (1997): Kleine Philosophie der Naturwissenschaften. (Beck) München.

Mettam, C. (1971): Functional design and evolution of the polychaete *Aphrodite aculeata*. J. Zool., London **163**: 489–514.

Trueman, E. R. (1968): Burrowing Habit and the Early Evolution of Body Cavities. Nature **218**: 96–98.

Trueman, E. R. (1975): The Locomotion of Soft-Bodied Animals. (Edward Arnold Publ.) London.

Trueman, E. R. (1980): Swiming by jet propulsion. In: Elder, H. Y. & Trueman, E. R. (Eds.): Aspects of Animal Movement. (Cambridge University Press) Cambridge: 93–105.

Weingarten, M. (1992): Organismuslehre und Evolutionstheorie. (Kovac) Hamburg.

Weingarten, M. (1993): Organismen – Objekte oder Subjekte der Evolution? (Wissenschaftliche Buchgesellschaft) Darmstadt.

Similarities and differences:
The distinctive approaches of systematics and comparative anatomy towards homology and analogy

Dominique G. Homberger

Introduction

By the time Darwin published his "Origin of Species", both macrotaxonomy (i.e., the systematics and classification of higher taxa) and comparative anatomy were well established biological disciplines and had a long history of trying to understand the order and patterns that are apparent in Nature and her organisms. Both disciplines also shared a primary interest in structure and form, and both relied on comparative observations as a basis for their research. The many similarities between macrotaxonomy and comparative anatomy tend, however, to obscure the fact that these two disciplines have pursued very different research programs and, as a result, have dealt differently with the challenge that the Theory of Evolution represented for their theoretical and methodological frameworks. These differences resulted also in different usages of certain concepts, such as those of homology and analogy.

The present paper addresses some confusions that have obscured various issues in contemporary evolutionary biology by sketching the historical use of comparative observations and concepts in macrotaxonomy and comparative anatomy.

The "comparative method"

The act of comparing objects is a quintessentially human activity and has been practised probably long before it was formalized by Aristotle. Comparisons involve, at their most basic level, decisions about the degrees of similarity and difference among individual objects. The information gathered from such comparisons can, however, be used for different ends and subjected to different interpretations.

One of the earliest and most generally applied usages of comparative data is the process of classifying objects, be they inanimate, biotic, or human-made (Mayr 1997). At the most basic level of the "classificatory comparative method", objects are classified by grouping them according to their similarities and differences in particular aspects of their form or function. The criteria according to which objects are classified are usually quite pragmatic and utilitarian and follow considerations of efficient information storage and retrieval (Mayr 1982, 1997; Janich 1992). An exception to this approach – and for biology the relevant one – is the classification of natural objects, for which a "natural" order or pattern has been perceived since the inception of philosophical thought, but for which classificatory criteria are not as obvious as for human-made objects or for natural objects that are used for specific purposes, as in pharmacy and technology (Janich 1992). It is this aspect of comparative work that is the domain of naturalists and systematists involved in macrotaxonomy.

Another usage of comparative data is the process of juxtaposing specially selected objects with particular combinations of structural or functional properties to gain insights into specific questions, as in the application of what is commonly called the "comparative method" and the "experimental method". As Mayr (1982) has already shown, the "comparative method" is not the opposite of the "experimental method", because both methods share many aspects, foremost among them being the basic process of comparison. In a typical experiment, an object is manipulated to make it differ in a particular aspect from its original, non-manipulated state. A subsequent comparison of the altered and original states of the object is expected to provide insights into the function of a particular aspect of the object. This "experimental comparative method" is especially powerful for sciences, such as physiology, that ask "How?" questions (Mayr 1997), or, in other words, that are concerned with function in its broadest sense. In a "natural experiment", which is the setting used by the "comparative method" the compared objects are not, or cannot be, manipulated, but must be selected judiciously so as to differ only in a particular aspect (see, e.g., Homberger 1988). A subsequent comparison of such objects can provide various insights, depending on whether the compared objects differ in a structural or in a functional aspect. This "natural comparative method" using natural experiments is especially powerful for sciences, such as comparative anatomy, evolutionary biology, astronomy, linguistics, and sociology, that ask "Why?" questions (Mayr 1997), or, in other words, that deal with objects having a historical component.

Thus, we can distinguish at least three usages of comparative data in biology: (a) the *classificatory comparative method* used by macrotaxonomists; (b) the *experimental comparative method* used by physiologists; and (c) the *natural comparative method* used by comparative anatomists. All three methods are special applications of the *comparative method*.

The pre-Darwinian roots of macrotaxonomy

The goal of systematics and macrotaxonomy is to organize the great diversity of the biotic part of Nature so as to facilitate information storage and retrieval (Mayr 1997). This is achieved through a series of comparisons between different organisms (see also Strickland 1840a: 189). Similar organisms are placed in the same group, dissimilar ones in separate groups. Eventually, a classification emerges that consists of hierarchically arranged groups of organisms that are connected to other groups by various degrees of similarity and separated from them by various degrees of dissimilarity. However, already Aristotle realized that the concept of similarity among organisms is not a straightforward one. For example, he recognized that the "fish-like" construction that is common to fishes and cetacean mammals is not a "good" criterion of similarity, but that "mammal-like" features, such as vivipary, hair, lungs, and mammary glands, are "good" criteria of similarity (see Cole 1949). As the not uncommon agreement among classifications proposed by scientists as well as by non-western or scientifically naive people (Mayr 1997) shows, the distinction between "good" and "wrong" criteria of similarity can often be made readily and intuitively (see also Strickland 1840a: 184–185). Nevertheless, the myriads of classifications that have been proposed since Aristotle also show that the distinction between "good" and "wrong" criteria of similarity is often open to interpretation (see also Strickland 1846a: 355). The reason for this uncertainty, at least until the publication of Darwin's theory, was that the organizing principle of Nature was not known.

Nevertheless, by the time Darwin was to publish his Theory of Evolution, systematists (in this paper used in the sense of biologists who are interested primarily in macrotaxonomy, or the classification of supraspecific taxa) had worked out a methodology for establishing a "natural system" that would reflect the "true" organizing principle of the biotic part of the world and that was based on an hierarchical arrangement of similarities among organisms and taxa. This methodology, which was compiled by Strickland (1840a, b, 1846a, b), was based on specific guidelines for the distinction between "true", or "essential", similarities, which would lead to the natural, "correct" system, and "accidental" similarities, which would lead to an artificial, "wrong" system (see Strickland 1840a, 1846a, b). Hand in hand with this formalization went a definition of concepts, such as natural system, affinity, homology, and analogy (see Table 1, p. 70).

According to this systematic methodology, affinities, or homologies, represent structural similarities among organisms and may have either similar or different functions. [As Bock & von Wahlert (1965) showed, the concept of "function" comprised until recently the gamut of functional aspects of organisms from physical properties to the interactions of their organs with the environment]. "Wrong" similarities became known as accidental similarities, or analogies, and were similarities that involved both structures and their functions. Thus, analogies, in contrast to homologies, always had the same function. The difference between analogous structures on the one hand and homologous structures with similar functions on the other hand was that the former had "a reference to the laws and properties of external and often inorganic matter" (Strickland 1846a: 360). In other words, analogies could be explained on the basis of physical laws; they were adapted to their environment and could be explained as a response to its physical requirements. Homologies, however, were "produced in conformity with the laws of the organic Creation" (Strickland 1846a: 360). In other words, they were the result of Divine Laws, or the laws of Nature, underlying the Natural System. Divine Laws are inherently unknowable, but were thought to be perceivable by trying to uncover the patterns and regularities that had been created by them. From these patterns, generalizations, or empirical principles, could be extracted, which in turn could be applied to new sets of comparisons for the proper identification of essential similarities, or homologies.

One of these principles was the "principle of the subordination of characters", which is attributed to Linné and Cuvier by Strickland (1846a), and which "teaches us to give to each point of structure its due weight, and to attach more value to those peculiarities whose immediate influence on the mysteries of life often renders them the most difficult for our senses to appreciate, than to those external characters which, though most conspicuous to the eye, are but remotely connected with the real Essence of the creature" (Strickland 1846a: 355). Thus, by following this principle, the presence of a nervous system, for example, provides one of the most important and basic criteria for essential similarity among organisms; because it is common to vertebrates and insects, it is also a criterion, or character, of "wide affinity". In contrast, the presence of a black plumage or pelage does not provide support for essential similarity between a crow, a black bear, and a European mole. In other words, the external character "black" is an accidental or analogous similarity among these animals. It may, however, be a criterion of homologous similarity between two species of crows and is, therefore, a criterion of "close affinity".

Law-like principles, such as the one described above, were reflections of the Divine Laws, or Laws of Nature, underlying the Natural Classification in the same manner as utilitarian considerations underlie the classification of human-made objects. Because these principles were based on observations of the actual biotic part of Nature, their application often resulted in a system that was remarkably "natural" or, as we would say today, "evolutionary", even before the Theory of Evolution had been proposed.

Finally, since, according to the systematic methodology, function cannot be used as a criterion for identifying essential similarities, and since analogous similarities are useless for establishing a natural system, function as a property of features can be disregarded in macrotaxonomy. In the words of Strickland (1846a: 359): "…although it is generally true that certain organs are destined to perform certain definite functions, yet the exceptions are so frequent as to make us attach a minor degree of importance to *function*, while we give the fullest weight to those essential properties which form the only test of real affinity." (Emphasis in original).

The post-Darwinian developments of macrotaxonomy

Contrary to a generally held belief, the established methods in systematics were not questioned or changed under the impact of Darwin's publication of his theory (see also Remane 1952: 11–13). Rather, they remained in use and were elaborated further over the years, while the established terms and concepts were "umfunktioniert", which can be roughly translated into English as "adjusted to serve new purposes" (see Table 1). An intriguing question is why a theory that had a fundamental impact on virtually every aspect of the image that humankind had formed about itself, the world, and the universe, had so little influence on the very science that had been trying to understand the essence of Nature.

Actually, two main reasons were responsible for this apparent paradox. First, the factual contents of Darwin's book did not challenge the classifications resulting from the efforts of systematists. On the contrary, many arguments that Darwin used to support the various parts of his theory were based on examples taken from studies created by naturalists and systematists. Furthermore, the genealogical relationships that were proposed by Darwin did not differ substantially from the Natural System, the reason for which was alluded to earlier in this paper. Thus, there was no immediate impetus for systematists to reevaluate their methods and force a paradigm shift in macrotaxonomy.

Second, the major and truly revolutionary tenets of Darwin's theory were of no obvious relevance to the established research program of systematists. For example, the definition of evolution as "descent with *modification*" (my emphasis) was probably of little interest and import for a research program that was focused on the establishment and ranking of *similarities* among organisms. Also, the concept of natural selection as a mechanism explaining the phenomenon of adaptation was of little relevance to a research program that had decided to disregard functional aspects of organisms. Therefore, the pre-Darwinian received wisdom and Darwin's theory could be compartmentalized as the practice of systematics on the one hand and the theory of systematics on the other hand.

This compartmentalization has determined to a great degree the development of post-Darwinian macrotaxonomy and has persisted to our present days. Three main trends can be recognized. First, efforts have been concentrated on finding ever "better" characters that could reveal the natural system (recast in various evolutionary terms after 1859) more easily and with fewer errors. Examples of such "better" characters have included egg-white proteins, DNA, mitochondrial DNA, mitochondrial cytochrome-b gene, etc. Second, major efforts have been spent in refining established empirical rules and principles (most of which are appropriated from non-biological disciplines, such as logic, mathematics, and philosophy) and in creating new ones to facilitate and codify the use of similarities to create a hierarchical system of groups (e.g., phenetics, cladistics or phylogenetic systematics), to translate hierarchical classification schemes into historical accounts (e.g., outgroup comparison, use of the parallelism between evolutionary transformations and developmental processes), and to resolve contradictory results (e.g., parsimony, various computer programs to reach consensus results). Third, there has been a continued general disinterest in the concept of non-homology as well as in concepts relating to functional aspects of organisms (e.g., analogy, adaptation) and in their significance for the practice and theory of macrotaxonomy [an exception is evolutionary systematics or taxonomy (Mayr 1982, 1997), e.g., Homberger (1980); for rare recent examples, see McCracken & Sheldon (1997) and Naylor & Brown (1997)].

The pre-Darwinian roots of comparative anatomy

Comparative anatomy shares Aristotelian roots with macrotaxonomy, but already differs in its research program at this early stage by its focus on the structure of organs that lie under the surface of organisms and need to be revealed by the use of dissection as a research method. Foreshadowing post-Renaissance comparative anatomy, Aristotle was interested in the function of organs as well as in analogies per se, that is, in the phenomenon of different organs fulfilling the same function in different organisms, as, for example, his meticulous description and functional analysis of a mammal-like placenta in a shark seem to indicate (cited after Cole 1949: 34–35).

Half a millennium after Aristotle, Galen also dissected and described the anatomy of a great variety of animals, but the assumptions underlying his comparative approach appear to have been different from those of Aristotle as well as from those of comparative anatomists after the Renaissance. Galen did not revel in the diversity of the natural world as Aristotle did, but assumed that all (vertebrate) animals, including human beings, had essentially the same basic anatomy and that they were composed of basically the same structures. Differences in the appearance of structures and organs among different animals were ascribed to differences in their relative proportions (O'Malley & Saunders 1983: 28). In keeping with these assumptions, Galen generally did not identify the animals he had dissected, apparently because this information was not relevant for the description of *the* anatomy of (vertebrate) animals and human beings. According to Cosans (in press), Galen suggested that the anatomical similarity between different species "is evident and clear proof that it is through the operation of a single source of wisdom, which concerned itself with them, that all the bodily parts of animals have been built up and created". However, as his numerous contributions on the workings of the human body show, Galen had, like Aristotle, and true to his profession as physician, an abiding interest in the functional aspects of structures (Cosans 1997, in press). Thus, two of the defining conceptual characteristics of comparative anatomy, namely the concept of a basic *bauplan* that is common to all animals and the primary interest in the function of structures, can be traced back to Galen.

These Galenic concepts persisted well into the Renaissance. Leonardo's (1452–1516) composite drawings, in which organs and body parts of animals and human beings were combined, reveal his understanding that organisms are built according to a basic construction plan despite the obvious structural differences among them (O'Malley & Saunders 1983: 28), and his studies in biomechanics clearly indicate his interest in function. However, the Renaissance probably marked also the beginnings of a growing realization of an actual diversity of structures among animals. Without this realization, it would be difficult to imagine how Leonardo could have prepared drawings of great accuracy of a bear's foot and a bird's wing (shown in O'Malley & Saunders 1983: 204–213).

Belon's (1555) famous juxtaposition of a human and an avian skeleton (reproduced in Panchen 1994) most probably represents a further development of a comparative anatomist's interest in structural differences that are superimposed on the received and reconfirmed knowledge of a basic common construction plan in animals. It seems obvious that Belon compared two *very* different organisms to show that they, nevertheless, shared commonalities. Thus, it would be inappropriate to equate Belon's use of the words "affinité" and "prochain" to describe the similarities between birds and human beings with the words "affinity" as used by Strickland (1840b) and with the direct translation "closely related", respectively, and, thus, to imply that Belon's similarities were the equivalent of today's homologies, as is frequently done (e.g., Hall 1994: 4; Panchen 1994: 43). Belon's comparison of a bird and a human being was not done for the purpose of classification and, therefore, it was not necessary for him to distinguish between different kinds of similarities as it was for macrotaxonomists. Such a reinterpretation of Belon's famous drawing also provides a key for understanding the puzzling fact that comparative anatomists – from François Hérissant to Richard Owen – have used various terms (e.g., homologies, analogies, correspondence, affinities, similarities, etc.) indiscriminately and interchangeably when describing structural similarities among different organisms.

Post-Renaissance comparative anatomists became increasingly fascinated by the differences among organisms while maintaining their primary interest in the function of structures. Thus, they contrasted very different animals (see Table 2), and many of the structural and functional similarities they nevertheless discovered were, in today's terminology, homoplasies. This aspect of the comparative anatomists' research program is well captured by Hérissant (1757: 279): "Le premier objet de l'Anatomie, & le plus intéressant pour nous, est la connoissance des parties qui entrent dans la composition du corps humain: elle nous fournit souvent des lumières par rapport à cet objet important, lors même qu'on étudie les parties intérieures des animaux dont la structure semble s'éloigner le plus des nôtres. Cette dernière étude, qu'on nomme l'*Anatomie comparée*, a toûjours au moins des faits extrêmement curieux à nous apprendre, lors-qu'elle nous fait voir combien diffèrent entr'eux les organes que l'Auteur de la Nature a employés dans différens animaux pour parvenir à une même fin, pour produire des effets assez semblables." [The first objective in anatomy – and for us the more interesting one – is the knowledge of the parts that make up the human body. This knowledge often provides us with insights into this important subject – human anatomy – even when we study the inner organs of animals whose structure seems to differ the most from ours. This type of study, which we call comparative anatomy, has always to teach us at least some extremely intriguing facts when it makes us realize how different the organs are which the Creator of Nature has used in different animals to reach the same goal, i.e., to generate rather similar effects. – my translation].

At the latest by the first half of the 19th century, the theoretical underpinnings of the "natural comparative method" (see earlier in this paper) had been elaborated to create conceptual and methodological connections among the various concepts of similarity, difference, function, and the common constructional plan of animals. Despite these elaborations, the concept of similarity remained a single entity, or, in other words, the need to distinguish between two different kinds of similarity (i.e., homologous and analogous ones) did not arise, at least not until Darwin's theory was published. Duvernoy (1830), for example, described the conceptual connections thus: "Quel est le mécanisme qui opère ces grands movemens? ne sont-ils dus simplement qu'à une extension des moyens ordinaires, employés dans les autres animaux de chaque classe? ou la nature a-t-elle été forcée d'en créer de nouveaux? Je dis à dessein *forcée*, parce qu'elle ne semble presque jamais recourir à d'autres expédiens et construire sur un nouveau plan, que lorsqu'il lui a été impossible de suivre son premier modèle. Ces questions ne tiennent pas simplement à l'explication de phénomènes particuliers à certains animaux; elles sont encore liées, comme l'on voit, aux lois générales de l'organisation, et doivent intéresser, sous ces différens points de vue, le physiologiste autant que le naturaliste." [What is the mechanism that operates these large movements {of the tongue of hummingbirds and chameleons}? Are they not simply extensions of the regular means used by other animals in each class? Or was nature forced to create new ones? I intentionally say "forced" because she {nature} seems almost never to resort to new mechanisms and to build according to a new plan, except when it has been impossible for her to follow her original model {plan, blueprint}. These questions do not relate only to the explanation of particular phenomena in certain animals; they are also tied, as one can see, to general laws of organisation, and {thus} have to be of interest, under these various points of view, to the physiologist {i.e., anatomist interested in function} as much as to the naturalist {i.e., systematist}. – my translation].

A few years later, Duvernoy (1836) made explicit the connections between the various concepts and the "natural comparative method": "En multipliant les comparaisons, en appréciant non-seulement les différences les plus remarquables, mais encore celles qui le sont moins, on arrive peu à peu à reconnaître les ressemblances générales, et à juger ce que chaque organe a de constant, d'essentiel, pour le constituer, et qui le distingue des circonstances organiques qui ne font que le modifier, qui le perfectionnent ou le détériorent, afin de le mettre en harmonie, selon les besoins de l'existence, avec l'ensemble de l'organisme. L'on parvient ainsi à l'autre but de cette étude, celui de découvrir le plan commun d'organisation de certains groupes; celui encore qui doit fournir des matériaux plus ou moins importans à la physiologie générale." [By increasing the number of comparisons, by paying attention not only to the most noticeable differences, but also to those that are less so, one succeeds gradually in recognizing general similarities and in evaluating what each organ has that is constant and essential, which constitutes it {the organ} and which separates it from the organic conditions which only modify it – which improve or reduce it – in order to put it in harmony, according to the needs of existence, with the whole of the organism. In this manner, one reaches also the other goal of {such a} study, namely to discover the constructional plan common to certain groups, the one which also has to provide the means and materials that are more or less important for the general physiology {functioning}. – my translation].

Comparative anatomical studies are typically based, explicitly or implicitly, on "natural experiments", in which objects are juxtaposed in such a manner that they differ or correspond in specific aspects (see Table 2). Owen (1837: 34) provides a particularly explicit example of a rationale for such a natural experiment: "... I... have dissected the olfactory nerve in [the *Vultur aura* or Turkey Buzzard]; as also in a Turkey which seemed to me to be a good comparison, being of the same size, and one in which the olfactory sense may be supposed to be as low as in the Vulture, on the supposition that this bird is as independent of assistance from smell in finding his food as the experiments of Audubon appear to show."

Thus, the research program of comparative anatomists differs fundamentally from that of macrotaxonomists and is characterized by (a) an equal interest in structure and function to understand the handiwork of the Creator or the engineering capabilities of Nature; (b) an equal interest in structural and functional differences and similarities to uncover a basic constructional plan (i.e., "unité de plan", "bauplan", "archetype", etc.); and (c) a clear methodology of comparison (see also Table 3). This means, that for comparative anatomists any similarities among animals were of interest, irrespective of whether they represented, in today's terminology, homologies or homoplasies, provided that they existed despite differences and, therefore, indicated the existence of a basic constructional plan.

Systematists, who had been most concerned with the different concepts of structural similarity, interpreted the undifferentiated concept of structural similarity, which was used by comparative anatomists, as a major flaw in the comparative anatomical approach, as is well illustrated by Strickland's (1846a: 362) comment: "It would be an improvement in the language of Comparative Anatomy, if the term *analogous organs* were limited to the sense above defined. The serrations in the beak of a duck, for instance, are *analogous* in form and in function to *teeth*, but in their essential nature they are only a corneous modification of the *lips*. Most anatomists, however, would habitually say that the beak of a bird is *analogous* to the lips of a Mammal, though it must be evident how much more precise their language would become if they spoke of this *essential* relation as an *affinity*, and applied the word *analogous* to *formal* or *functional* relations only." (Emphases in original)

Richard Owen, who was a comparative anatomist by training and inclination, but interacted closely with naturalists and systematists professionally (Rupke 1992, 1994), must have been motivated by such criticisms, which did not originate just with Strickland, to include a definition for the terms of analogy and homology in his later papers (Owen 1843, 1848; fide Panchen 1994): "Homologue: The same organ in different animals under every variety of form and function. Analogue: A part or organ in one animal which has the same function as another part or organ in a different animal." These definitions are not equivalent to the definitions of systematists, who are interested only in structural similarities when trying to establish a natural system and are not interested in the differences *per se* among animals. Thus, for systematists, similarity in structure is a *conditio sine qua non* for both homologies and analogies (see Table 1), while differences in structure or function are irrelevant for their approach and are neither categorized nor given special comparative terms. Owen's definitions, in contrast, allow for both homologous and analogous organs to differ in structure and reflect the comparative anatomists' understanding of function being a crucial aspect of form and of both form and structure being variable (see Table 2). Strickland (1846a: 358) himself seems to have been quite aware at least of the fundamentally different concepts of analogy in comparative anatomy and macrotaxonomy: "In works of comparative anatomy it is customary to speak of those members which are essentially equivalent in two organic beings as *analogous* organs, but we shall soon see that the word analogy has a very different sense…". Fürbringer's (1873: 238) criticism of a study published in 1868 illustrates this pre-Darwinian comparative anatomical concept of analogy succinctly: "Der Verfasser nimmt den von früheren vergleichenden Anatomen in richtiger Einsicht verlassenen alten Standpunct der Analogien ein, wonach die functionelle Bedeutung der Muskeln als Vergleichungspunct benutzt wird…" [The author adopts the standpoint of analogies, which has quite rightly been abandoned by earlier comparative anatomists, and according to which the functional significance of muscles is used as a point of reference for comparisons… – my translation].

Owen evidently tried to understand and adopt the macrotaxonomists' distinction of two different concepts of similarity, but in the process of doing so, he reworded the macrotaxonomists' definitions to fit his own comparative anatomical conceptions. Clearly, the macrotaxonomists' proposition to consider function as an "unreliable" aspect of an organ, which should, therefore, be disregarded, could not have made sense to a comparative anatomist's mind set. Furthermore, for comparative anatomists, analogous organs need not have a similar structure, as some of their studies clearly demonstrate (see Table 3). Irony reached one of its pinnacles in the history of science when, according to Strickland (1846b), Owen claimed priority in having introduced the concept of homology into comparative anatomy and when Strickland (1846b) praised Owen for doing so, since both men seem to have been blissfully unaware of the giant misunderstanding between them. One can see here the beginnings of the "comedy of errors" that has bedeviled the discourse on homologies and analogies between biologists of differing training and inclinations ever since. That this confusion of concepts is not just of academic interest, but that it fundamentally affects the biological interpretation of observations, is exemplified by the different conclusions that were reached by Owen, a comparative anatomist, and Milne-Edwards, a macrotaxonomist, on the subject of lungfishes, with Owen reaching the conclusion that lungfishes are fishes and Milne-Edwards concluding that they are "reptiles" (i.e., amphibians) (see Milne-Edwards 1841).

The comparative anatomists' specific approach to the concepts of function, similarity, and difference lies also at the heart of two major concepts, namely the idea of the existence of a basic constructional plan and the idea that both structures and functions are variable and, thus, can change. The Galenic concept of a basic constructional plan underlying animal form runs through the historical development of comparative anatomy like a red thread, picking up different names and being fitted into different contexts along the way, but never being abandoned. From this vantage point, the "idealistic morphologists" and "German Naturphilosophen" should be rehabilitated as legitimate participants in the scientific developments preparing the grounds for Darwin's theory (see also Remane 1952: 14).

Pre-Darwinian ideas of structural and functional change (e.g., transmutation, transcendentalism, transformationism, etc.) were developed almost exclusively by comparative anatomists and not by systematists, whose preoccupation with classification are likely to have inhibited thoughts of fluidity, change and intermediary stages between organisms (see also England 1997: 269). Furthermore, the simultaneous preoccupation of comparative anatomists with both commonalities and differences among animals logically raised questions about the reasons for the existence of particular structures and functions (see also Nyhart 1995: 39–42). As we shall see, both the concepts of a common constructional plan and of structural and functional change predisposed at least some comparative anatomists to incorporate and integrate Darwin's theory more completely than did macrotaxonomists.

The post-Darwinian developments of comparative anatomy

The integration of Darwin's theory into the research program of comparative anatomy did not proceed as directly as is sometimes portrayed (e.g., Russell 1916: 247; Remane 1952: 13–14; Mayr 1997: 71 & 180). Darwin's theory of "descent with modification" did, however, provide a causal explanation for the existence of *both* the differences and similarities among organisms and, thus, addressed a fundamental part of the research program of comparative anatomists. In particular, the concept of natural selection provided an explanation for the adaptive nature of homoplasies (in today's terminology) and for the existence of differences among animals with a similar basic constructional plan. Darwin's Theory, however, also forced comparative anatomists, at last, to develop criteria to distinguish two different kinds of similarity, namely the one due to common descent (i.e., homology) and the other due to similar selective environmental agents (i.e., analogy or homoplasy). Furthermore, comparative anatomists now also had to find criteria to identify homologous features that had been modified by diverging evolution to such a degree that they were no longer "similar".

The two men who are credited with integrating Darwin's theory into comparative anatomy, namely Ernst Haeckel and Carl Gegenbaur (Russell 1916, Nyhart 1995), actually took very different approaches. Ernst Haeckel, who was a naturalist and systematist by inclination and training, synthesized pre-Darwinian concepts from macrotaxonomy and comparative anatomy and reformulated them in evolutionary terms. Examples of pre-Darwinian contributions to his morphological theories are the various empirical analogies (e.g., ontogeny recapitulates phylogeny) that were supposed to provide a timeline in a series of comparisons or to permit the reconstruction of ancestral forms, which were remarkably similar to pre-Darwinian "archetypes".

In contrast, Carl Gegenbaur, a comparative anatomist by inclination and training, seems to have recognized what has escaped most biologists, namely that Darwin's theory was not simply a new scientific theory, but that it contained a core of concepts and underlying methods that had been adopted from the social and historical sciences. One of the crucial insights that seeded the theory of evolution in Darwin's mind was the recognition of an analogy between the origin and change over time of rocks, the historical development of languages, and the origin of species (Desmond & Moore 1991: 215–216; Homberger 1998; see also Darwin's quotes in M. Gutmann 1995: 78 & 82). The close connection between Darwin's theory and history has been appreciated more fully in recent years (e.g., Allen 1992, Hull 1992, Nitecki 1992, Richards 1992, Mayr 1997: 37 & 277), but Gegenbaur and his associates were already quite aware of it, as can be shown by examples from their works: "Was dem Historiker die Geschichtsquellen sind, das müssen dem Anatomen die anatomischen Thatsachen sein, eben so sicher festgestellt, ebenso vollständig, aber ebenso nur den Ausgangspunct zu weiteren Folgerungen bildend." (Gegenbaur 1875: 4) [What the historical sources are for the historian, the anatomical facts have to be for the anatomist, observed with equal certainty, equally complete, but also forming only the starting point for further inferences and conclusions. – my translation] and "Wenn, wie in so hohem Grade wahrscheinlich gemacht worden die organischen Formen nicht unverändert von Anfang an bestehen, sondern durch langsame Umwandlung aus einander hervorgegangen, so musste es wohl unsere nächste Aufgabe sein diese Umwandlungen zu verfolgen. Wir hatten hiermit das Recht erworben auf die Gegenstände unserer Forschung dieselben *historischen Methoden* anzuwenden, die auch andere Gebiete, wo eine Entwickelung längst anerkannt worden, beherrschen: so die Geschichte unserer politischen und geistigen Entwickelung, so die vergleichende Sprachforschung." (Strasburger 1875: 57) [If, as was made probable to such a high degree, organisms do not exist unchanged from the beginning, but have arisen from each other through slow transformation, our next task had to be to follow (study) this transformation. Thus, we have acquired the right to apply to our research objects the same methods that dominate other disciplines which have recognized a historical development for a long time, such as the history of our political and cultural development and comparative linguistics. — my translation]. Evolutionary change from one condition of an organ or body part to the other was reconstructed by using a variety of principles that were based on functional-morphological, bioconstructional and biological-ecological observations to restrict the infinite number of theoretically possible historical pathways to those that are consistent with biological facts and with the theoretical framework in biology, i.e., with the theory of evolution.

The genuinely historical approach of Gegenbaur's school (see, e.g., Fürbringer 1873: 237–344; Gegenbaur 1875, 1889) seems not to have been recognized previously as such, possibly because historical methods, being based on comparative data, may have been mistaken for the comparative methods used in pre-Darwinian times, especially since several pre-Darwinian terms were retained. Furthermore, the studies of the Gegenbaur school were largely forgotten as the medical schools started to concentrate on human anatomy and abandoned the teaching of comparative anatomy, and as the "modern" experimental biological disciplines (e.g., genetics, Entwicklungsmechanik, ecology) came to dominate the field of zoology (Nyhart 1995).

The cross-pollination of macrotaxonomy and comparative anatomy

Until well into the middle of the 19th century, it was widely recognized that macrotaxonomy and comparative anatomy were two separate biological disciplines with distinct research programs (see Table 4), as can be demonstrated by a few selected quotes. "I must not pass without notice another branch of our science of the deepest interest and highest importance, and more particularly as we have to lament that hitherto it has been very imperfectly cultivated…in these islands [Britain], I mean *the Comparative Anatomy* of animals" (Kirby 1823: 462). "…for it is no more expected that systematists should have already unravelled all the resemblances between species contemplated by the Creator, than that anatomists should have arrived at the final cause of every organ of the human body" (Strickland 1846b: 221). "That the opportunities of acquiring a knowledge of the organization of the extinct Bird once inhabiting the island of Mauritius should be now irrevocably past, is… a subject of the deepest regret to every one interested in the advancement of zoological science: whether he be engaged as a systematic naturalist in unravelling the intricacies of the natural system; …or as a philosophical anatomist, in investigating the principles which regulate the deviations from a typical standard of organization, …" (Owen 1841).

Towards the end of the 19th century, however, the two scientific disciplines had started to become less distinct. One of the reasons, was, as we have seen, that comparative anatomists had to incorporate certain concepts from macrotaxonomy, such as the distinction between homologous and analogous similarities, into their conceptual and theoretical framework in order to incorporate Darwin's theory into their research program. Macrotaxonomists, in their continual search for criteria that could be used for classification, integrated anatomical characters into their analyses. Comparative anatomists, in turn, increasingly used their results also to discuss the systematic positions of the organisms they studied. The history of this cross-pollination between systematics and comparative anatomy, as well as the role of post-Darwinian paleontology in these developments, has yet to be reconstructed and will only be alluded to in this paper with the help of a few selected examples.

The efforts of Richard Owen and Ernst Haeckel in trying to synthesize the concepts developed by macrotaxonomists and comparative anatomists have already been mentioned earlier in this paper.

With the eclipse of comparative anatomy as a part of human anatomy in medical schools, some comparative anatomical concepts and research were kept alive originally by scientists that had been trained as comparative anatomists, but had found employment in academic departments of biology in the broad sense (Nyhart 1995). It was here that the cross-pollination between systematics and comparative anatomy became most effective and led to today's impression of indistinct boundaries between the two disciplines.

Evolutionary systematics (Mayr 1997) and evolutionary morphology is a product of such an amalgamation of pre-Darwinian systematics and comparative anatomy with the post-Darwinian disciplines of genetics, biogeography, ecology, and the synthetic theory of evolution.

Remane (1952) and his codification and formalization of clear criteria (Homologiekriterien) for the identification of homologous organs that have diverged in structure and, thus, cannot be identified through the traditional criterion of structural similarity (see Remane 1952: 31) represents another example of a product from such a cross-pollination. Remane's (1952) criteria for homology and for the establishment of a time line are based on empirical principles that are not causally connected to the theory of evolution, but are still influential and frequently used in comparative morphological and other comparative studies, especially in the German-speaking scientific community. Remane's (1952) work has much in common with that of his compatriot Hennig (1950), who codified and formalized criteria for the identification and use of homologies that are similar in structure for classificatory purposes. Hennig's (1950) criteria are based mostly on empirical principles that concern patterns of distribution. The success and legacy of both Remane's (1952) and Hennig's (1950) work lie in the clarity and easy application of rules for the identification of homologies. The problem with both approaches is that they do not genuinely reconstruct historical processes and entities, as they do not use physical, biological and evolutionary mechanisms and processes as constraints for time-dependent change as would be required for a true post-Darwinian comparative biology. Both Hennig's and Remane's methods are essentially codifications of pre-Darwinian concepts and approaches.

In complete contrast stands the work of Wolfgang F. Gutmann and his co-workers (e.g., W.F. Gutmann 1981, 1989), who applied physical, biological and constructional principles in their rigorously historical reconstructions of evolutionary changes of organisms and, thus, stand in the tradition of the school of thought established by Carl Gegenbaur and his co-workers. In doing so, they may, however, not have challenged Darwinian dogmas ("altdarwinistische Dogmen"), as they claimed, but rather the pre-Darwinian concepts and approaches that have been kept alive to this day despite the "Darwinian Revolution" (see also Starck 1977).

Conclusions

Concepts and methods are developed and used as part of research programs. Thus, concepts reflect particular research programs and cannot be transferred tel quel to different research programs. Analyses of scientific methods and concepts need to consider this when attempting to evaluate the validity of various approaches to scientific problems.

Acknowledgements

The invitation by the late Wolfgang F. Gutmann to participate in the 7th International Senckenberg Conference provided me with the opportunity to present my ideas. Wolfgang F. Gutmann, by engaging me in discussions in his stimulating, uncompromising and intellectually clear manner, and by giving generously of his time and knowledge, also forced me to maintain my interest in theoretical, cultural and historical issues in animal morphology. Walter J. Bock kept me engaged in thinking about theoretical issues in systematics and evolution and provided me with some of the references that proved crucial for the coalescence of the ideas presented here. Walter J. Bock, Michael Gudo, Mathias Gutmann, Christine Hertler, Martha Hyde, John C. Larkin, and A. Ravi P. Rau provided feedback in more recent times. Roberta Ruiz at Interlibrary Borrowing of the Middleton Library at Louisiana State University provided me with the difficult-to-find necessary references. Karen Romagosa typed the manuscript. I thank all for their much appreciated contributions.

Literature

Allen, G.E. (1992): Evolution and history: History as science and science as history. Pp. 211–239 in History and Evolution (M.H. Nitecki and D.V. Nitecki, eds.). State University of New York Press, New York.

Barkow, H. (1829): Anatomisch-physiologische Untersuchungen, vorzüglich über das Schlagadersystem der Vögel. Archiv für Anatomie und Physiologie, **4**: 305–496 & plates 8–10.

Belon, P. (1555): L'histoire de la nature des oyseaux. Guillaume Cavellat, Paris.

Bock, W.J. & von Wahlert, G. (1965): Adaptation and the form-function complex. Evolution, **19**: 269–299.

Cole, F.J. (1949): A history of comparative anatomy from Aristotle to the eighteenth century. Macmillan Press, London. (Reprinted 1975 by Dover Publications, New York).

Cosans, C.E. (1997): Galen's critique of rationalist and empiricist anatomy. Journal of the History of Biology, **30**: 35–54.

Cosans, C.E. (In press): The experimental foundations of Galen's teleology. Studies in History and Philosophy of Science: in press.

Desmond, A. & Moore, J. (1991): Darwin. Warner Books, New York.

Duvernoy, G.L. (1830): De la langue, considérée comme organe de préhension des alimens. Mémoirs de la Société d'Histoire Naturelle de Strasbourg, **1**: 1–16 & 5 planches.

Duvernoy, G.L. (1836): Mémoire sur quelques particularités des organes de la déglutition de la classe des oiseaux et des reptiles, pour servir de suite à un premier mémoire sur la langue. Mémoirs de la Société d'Histoire Naturelle de Strasbourg, **2**: 1–24 & planches 1–5.

England, R. (1997): Natural selection before the *Origin*: Public reactions of some naturalists to the Darwin-Wallace papers (Thomas Boyd, Arthur Hussey, and Henry Baker Tristram). Journal of the History of Biology. **30** (2): 267–290.

Fürbringer, M. (1873): Zur vergleichenden Anatomie der Schultermuskeln. I. Theil. Jenaische Zeitschrift für Medizin und Naturwissenschaft, **7**: 237–320.

Gegenbaur, C. (1875): Die Stellung und Bedeutung der Morphologie. Gegenbaurs Morphologisches Jahrbuch, **1**: 1–19.

Gegenbaur, C. (1889): Ontogenie und Anatomie, in ihren Wechselbeziehungen betrachtet. Gegenbaurs Morphologisches Jahrbuch, **15**: 1–9.

Gutmann, M. (1995): Die Evolutionstheorie und ihr Gegenstand. Verlag für Wissenschaft und Bildung, Berlin.

Gutmann, W.F. (1989): Die Evolution hydraulischer Konstruktionen: Organismische Wandlung statt altdarwinistischer Anpassung. Verlag Waldeman Kramer, Frankfurt am Main.

Gutmann, W.F. & Bonik, K. (1981): Kritische Evolutionstheorie: Ein Beitrag zur Ueberwindung altarwinistischer Dogmen. Gerstenberg Verlag, Hildesheim.

Hall, B.K. (1994): Introduction. Pp. 1–19 in B. K. Hall (ed.): Homology: The hierarchical basis of comparative biology Academic Press, San Diego.

Hennig, W. (1950): Grundzüge einer Theorie der phylogenetischen Systematik. Deutscher Zentralverlag, Berlin.

Hérissant, F.D. (1752): Observations anatomiques sur les mouvemens du bec des oiseaux. Histoire de l'Académie Royale des Sciences 1748, volume 65 (avec les Mémoires de Mathématiques et de Physique, pour la même Année): 345–386 & planches 15–23.

Hérissant, F.D. (1757): Recherches sur les organes de la voix des quadrupèdes, et celle des oiseaux. Mémoires (de Mathematiques et de Physique tirés des Registres de) l'Académie Royale des Sciences, 1753: 279–295 & planches I & II.

Homberger, D.G. (1980): Funktionell-morphologische Untersuchungen zur Radiation der Ernährungs- und Trinkmethoden der Papageien (Psittaci). Bonner zoologische Monographien No. **13**: 1 – 192.

Homberger, D.G. (1988): Models and tests in functional morphology: The significance of description and integration. American Zoologist, **28**: 217–229.

Homberger, D.G. (1991): The evolutionary history of parrots and cockatoos: A model for evolution in the Australasian avifauna. Acta XX Congressus Internationalis Ornithologicus: 398 – 403.

Homberger, D.G. (1994): Ökomorphologie der rotschwänzigen Rabenkakadu-Arten (*Calyptorhynchus* spp.) in Australien: Beispiel einer multispektiven Biodiversitätsstudie als Grundlage für die Rekonstruktion der Evolutionsgeschichte einer Artengruppe. Pp. 425 – 434 in W.F. Gutmann & D.S. Peters (eds.): Morphologie und Evolution. Senckenberg-Buch **70**. Kramer, Frankfurt a.M.

Homberger, D.G. (1998): Was ist Biologie? Pp. 11–28 in A. Dally (ed.): "Was wissen Biologen schon vom Leben? – Die biologische Wissenschaft nach der molekular-biologischen Revolution". Loccumer Protokolle 14/97. Evangelische Akademie Loccum, Germany.

Hull, D.L. (1992): The particular-circumstance model of scientific explanation. Pp. 69–80 in M.H. Nitecki & D.V. Nitecki (eds.): History and Evolution. State University of New York Press, New York.

Janich, P. (1992): Grenzen der Naturwissenschaft. Verlag C.H. Beck, München.

Kirby, W. (1823): Address of the chairman read at the meeting of the Zoological Club of the Linnean Society held at the Society's House in Soho Square, Nov. 29, 1823. Philosophical Magazine and Journal, **62**: 457–464.

Mayer, A.E.J.C. (1852): Ueber den Bau des Organes der Stimme bei den Menschen, den Säugethieren und einigen grossen Vögeln. Novorum Actorum Academiae Caesareae Leopoldo-Carolinae Naturae Curiosorum, **23** (2): 659–766 & plates 62–89.

Mayr, E. (1982): The growth of biological thought. Belknap Press of Harvard University Press, Cambridge, Massachusetts.

Mayr, E. (1997): This is biology. The science of the living world. Belknap Press of Harvard University Press, Cambridge, Massachusetts.

McCracken, K.G. & Sheldon, F.H. (1997): Avian vocalizations and phylogenetic signal. Proceedings of the National Academy of Sciences of the USA, **94**: 3833–3836.

Milne-Edwards, A. (1841): On the natural affinities of the *Lepidosiren*; and on the differing opinions of Mr. Owen and M. Bischoff with regard to them. Annals and Magazine of Natural History, Series 1, **6** (39): 466–468.

Naylor, G.J.P. & Brown, W.M. (1997): Structural biology and phylogenetic estimation. Nature, **388**: 527–528.

Nitecki, M.H. (1992): History: La grande illusion. Pp. 3–15 in M.H. Nitecki & D.V. Nitecki (eds.): History and Evolution. State University of New York Press, New York.

Nyhart, L.K. (1995): Biology takes form: Animal morphology and the German universities, 1800–1900. University of Chicago Press, Chicago.

O'Malley, C.D. & Saunders, J.B. de C.M. (1953): Leonardo on the human body. Dover Publications, New York.

Owen, R. (1837): Letter to W. Yarrell [on the Turkey Vulture]. Proceedings of the Zoological Society, London, part 5: 34–35.

Owen, R. (1841): On the anatomy of the Southern Apteryx (*Apteryx australis*, Shaw). Transactions of the Zoological Society, London (1838), **3** (4): 257–301 & plates 48–55.

Owen, R. (1843): Lectures on comparative anatomy and physiology of the invertebrate animals, delivered at the Royal College of Surgeons, in 1843. Longman, Brown, Green, and Longmans, London.

Owen, R. (1848): On the archetype and homologies of the vertebrate skeleton. John van Voorst, London.

Panchen, A.L. (1994): Richard Owen and the concept of homology. Pp. 21–62 in B.K. Hall (ed.): Homology: The hierarchical basis of comparative biology. Academic Press, San Diego.

Remane, A. (1952): Die Grundlagen des natürlichen Systems, der vergleichenden Anatomie und der Phylogenetik. Akademische Verlagsgesellschaft Geest & Portig K.-G., Leipzig.

Richards, R.J. (1992): The structure of narrative explanation in history and biology. Pp. 19–53 in M.H. Nitecki & D.V. Nitecki (eds.): History and Evolution. State University of New York Press, New York.

Rupke, N.[A.] (1992): Richard Owen and the Victorian museum movement. Journal and Proceedings of the Royal Society of New South Wales, **125**: 133–136.

Rupke, N.A. (1994): Richard Owen, Victorian naturalist. Yale University Press, New Haven, London.

Russell, E.S. (1916): Form and function: A contribution to the history of animal morphology. John Murray, London. (Reprinted 1982 by University of Chicago Press, Chicago).

Starck, D. (1977): Tendenzen und Strömungen in der vergleichenden Anatomie der Wirbeltiere im 19. und 20. Jahrhundert. Natur und Museum, **107**: 93–102.

Strasburger, E. (1874): Ueber die Bedeutung phylogenetischer Methoden für die Erforschung lebender Wesen. Jenaische Zeitschrift für Naturwissenschaft, N.F., **8**: 56–80.

Strickland, H.E. (1840a): On the true method of discovering the Natural System in Zoology and Botany. Annals and Magazine of Natural History, series 1, **6** (36): 184–194.

Strickland, H.E. (1840b): Observations upon the affinities and analogies of organized beings. Magazine of Natural History, **4**: 219–226.

Strickland, H.E. (1846a): On the structural relations of organized beings. Philosophical Magazine and Journal of Science, **28**: 354–364.

Strickland, H.E. (1846b): On the use of the word homology in comparative anatomy. Philosophical Magazine and Journal of Science, **29**: 35.

Thuet, M.I. (1838): Disquisitiones anatomicae psittacorum. Dissertatio inauguralis Universitate Litteraria Turicense. Typis Orellii, Fuesslini et Sociorum, Turici.

Pre-Darwinian (e.g., Strickland 1840a, b, 1846a, b)	Post-Darwinian
Natural System (purportedly based on Divine Laws or laws of Nature)	Classification and Phylogeny (purportedly based on the Theory of Evolution)
Map of affinities {represented "in a pleasing manner [as] an artificial tree" (Strickland 1840: 190)}	Phylogenetic tree; cladogram
Essential (similar) characters (similar in structure, organic composition, and relative position; similar *or* different in superficial form and function)	Homologous characters (operational definition: similar in microstructure, relative position, and development; similar or different in external form and function) (theoretical definition: charcters, in two or more taxa, that are derived from the same one structure in the most recent common ancestor)
Analogous (similar) characters (non-essential, arbitrary, accidental; similar in structure and function because of influence of similar environments)	Homoplastic analogous characters (non-homologous; with similar structure and function in adaptation to similar environments)
Iconism (Strickland 1846a: 362) (similarity based on non-essential, similar structures with different functions)	Non-analogous homoplasy (similarity based on non-homologous, similar structures with different functions)
characters with wide affinities	primitive (plesiomorph) characters
characters with close affinities	derived (apomorph) characters

Table 1: Concepts in Systematics

Author	compared objects	external form	internal or detailed structure	function	similarities in today's terms
Belon (1555)	human being and bird	different (body shape)	similar (skeletal elements)	different or similar (depending on skeletal elements)	homology
Hérissant (1752)	jaw apparatus of human beings and birds	similar (as seen in intact heads)	different (skeletal elements)	similar (opening and closing mouth)	homology or analogy (depending on elements)
Hérissant (1757)	mammalian and avian voice apparatus (larynx and syrinx, respectively)	different	different	similar	non-homology, analogy
Barkow (1829)	arteries in birds and mammals	similar	different	similar	homology or analogy (depending on elements)
Duvernoy (1830, 1836)	tongue of mammals, reptiles, amphibians, and birds	different	different	similar (food acquisition)	homology or analogy (depending on elements)
Owen (1837)	olfactory nerves in turkey and turkey vulture	similar	different	similar	homology
Thuet (1838)	human and psittacine tongue	similar	different	similar (producing speech)	non-homology, analogy
Mayer (1852)	human, mammalian and avian larynx	similar	different	different (in terms of voice production)	homology

Table 2: Examples of classic comparative anatomical studies

Pre-Darwinian	Post-Darwinian
Looking for a reason ("Grund") for the observed differences and similarities among organisms	Using the Theory of Evolution to explain the observed differences and similarities among organisms
Finding correspondences among structures and organs of different organisms as evidence for a basic constructional plan	Reconstructing the evolutionary history of structures and organs
Using the comparative method to create natural experiments	Using natural experiments to test phylogenetic hypotheses
Dealing with both functional and structural aspects of the world	Dealing with various relevant biological disciplines and incorporating their data and concepts

Table 3: Features of the research program of comparative anatomy

Systematics	Comparative Anatomy
Similarities are focus of interest	Differences and similarities are of equal interest
Homologies are of primary interest	Homologies and analogies are of equal interest
Adaptive characters are accidental and due to environmental influences, and are of negligible interest	Adaptive characters exemplify handiwork of Creator or existence of natural selection, and are of greatest interest
Comparative method used to find similarities to create groups	Comparative method used to create natural experiments
Disregard of function hinders integration with Darwin's Theory	Emphasis on function facilitates integration with Darwin's Theory
Retention of pre-Darwinian methodology	Development and adaptation of new concepts and methods
Non-biological principles used to evaluate significance of similarities and differences and their patterns	Principles rooted in biology used to understand similarities and differences and their patterns
Main interest in evolution of organisms	Main interest in evolution of organs and structures

Table 4: Major differences between systematics and comparative anatomy

The organism as a necessary entity of evolution

RAPHAEL FALK

Hugo de Vries's "rediscovery", in 1900, of the rules of Mendelian inheritance was in reality a formalization of his reductionist view of development (Falk, 1995). The *unit character*, rather than the *organism*, was posited as the fundamental entity of biology. In his theory of *pangenesis* de Vries thus completed Weismann's conversion of Darwin's centripetal notion of inheritance and development into a centrifugal theory of the organism as a summation of essential units. "According to pangenesis the total character of a plant is built up of distinct units. These so-called elements of the species, or its elementary characters, are conceived of as tied to bearers of matter, a special form of material bearer corresponding to each individual character. Like chemical molecules, these elements have no transitional stages between them." (de Vries, 1900)

Wilhelm Johannsen's (1909) differentiation between the genotype and the phenotype, significant as it was for overcoming the de Vriesian notion of "unit character," was also a crucial step in the phasing-out of the organism as the elementary unit of systematics and evolution (see also Gudding, 1996). Both de Vries and Johannsen put forward theories of development that were based on essential entities that are (nearly) immutable, of which the individual organisms were mere ephemeral *epiphenomena*. Johannsen, in reality, was interested in the nature of systematics and "saw the genotype as the essence of the narrowest systematic unit, the elementary species" (Roll-Hansen, 1978, p. 221).

In the decades that followed, although "genes" and "genotypes" replaced "unit characters" and "pure-lines," respectively, heredity's essential nature was categorically juxtaposed against nurture, environment's inconsistency: Environment was conceived as the ever-present, unavoidable nuisance that unfortunately must always be reckoned with. Furthermore, the individual organisms, their development and function were epigenetic events that the hereditary essences displayed within an extended notion of the "environment". This was called by Johannsen the *genotype conception*: "The *personal qualities* of any individual organism do not at all cause the qualities of its offspring, but the qualities of both ancestor and descendent are in quite the same manner determined by the nature of ... the gametes – from which they have developed. Personal qualities are then *the reactions of the gametes* joining to form a zygote." (Johannsen, 1911, p. 130)

This *genocentric* notion was maintained and even elaborated further as long as the central issues of hereditary research were those of transmission and of population genetics. To the extent that the physiology and embryology of the operation of the genotype within the environment were mentioned they were relegated to technical complications, for which terms like "penetrance" – the proportion of individuals of a given genotype that show the expected phenotype, or "expressivity" – the degree of variability of the typical phenotype, were provided (Timoféeff-Ressovsky & Timoféeff-Ressovsky, 1926; Vogt, 1926). Other terms, like Mendel's own "dominance," were considered properties of the genes themselves, the evolution of which deserved attention (Fisher, 1928; Muller, 1950). Even the physiologically-minded Richard Goldschmidt related to the environmental impacts on organisms' development in a term like "phenocopy" (Goldschmidt, 1935), i.e., "a nonhereditary, phenotypic modification (caused by special environmental conditions) that mimics a similar phenotype caused by a gene mutation" (Rieger, Michaelis and Green, 1976). Drosophila geneticists used to denote for many years, next to the description of each mutation, its penetrance and expressivity (as well as its dominance) as indications of the "reliability" of that mutation in breeding experiments: Mutations with low penetrance or variable expressivity were too much subject to disturbing environmental fluctuations in their phenotype to be useful experimental material (see examples in Lindsley and Grell, 1968).

Jan Sapp noted that "Ironically, Johannsen's genotype/phenotype distinction offered geneticists the conceptual space or route by which they could bypass the organization of the cell, regulation of the internal and external environment of the organism, and the temporal and orderly sequence of development" (Sapp, 1987, p. 49). Perhaps even more ironic, though, is the fact that the reductionist bypass offered by the genotype/phenotype distinction, specifically designed to relieve biologists of appealing to irrational explanations like *vis vitalis* or *entelechy*, has systematically reintroduced irrational essentialism into biology only half a century after Darwin presumably eliminated it. Thus it transpired that the profound Darwinian notion of life as a unique, continuous, evolving, yet historically constrained phenomenon, was misconstrued. The organism, as an essential entity of the eternal dance of stability and change of the repeated trial and error in a world that is heterogeneous in space and time, was systematically ignored.

This caricature of the fading organism has been epitomized by Richard Dawkins's grotesque declaration that "We are all survival machines": "The evolutionary importance of the fact that genes control embryonic development is this: it means that genes are at least partly responsible for their own survival in the future, because their survival depends on the efficiency of the bodies in which they live and which they helped to build." (Dawkins, 1976, p. 25)

For Dawkins, the individual organism, its development and function, are simply the genes' way to make more genes, "One feature of this life in this world ... is that living matter comes in discrete packages called organisms.... It is legitimate to speak of adaptations as being 'for the benefit of' something, but that benefit is best not seen as the individual organism. It is a smaller unit which I call the active, germ-line replicator." (Dawkins, 1982, p. 4)

Johannsen's essentialist view of systematics was, however, challenged by Richard Woltereck (1909) already in the same year that he put forward his distinction between genotype and phenotype. Diagnostic quantitative traits of local strains, or *Elementararten* of *Daphnia* species from different lakes in Germany, which clearly formed pure lines, could be modified by changing environmental conditions. Although Woltereck's opposition to Johannsen's genotype/phenotype distinction was primarily based on his opposition to Johannsen's essentialist approach to systematics, by insisting on the role of environmental conditions on the determination of the systematic status of the species, he *per force* emphasized the role of the individual organism and its development in systematics. This way of thinking, however, obviously introduced epigenetic notions, which at the time, and for many years, branded Woltereck as "Lamarckian."

Woltereck described for each biotype its specific (phenotypic) response curve to environmental variations. It turned out that these response curves (for the mean values of a given variable) change for each biotype with every additional variable that was considered. Furthermore, there was no way of predicting the response-curve of one biotype from that of another (Figure 1). Woltereck recognized that the entities of life, whether species or biotypes, or individual organisms for that matter, were functions of complex relationships of environmental variability, not less than those of hereditary constancy and continuity. These inclusive relations may be denoted as the specific and relative *norm-of-reaction* (*Reaktionsnorm*) of the analyzed trait (Woltereck 1909, p. 135).

Although Woltereck's notion of the norm-of-reaction is of *traits* rather than that of the *organism*, it is in clear contrast to the reductionist genocentric attitude of most students of experimental heredity. The numerous studies in the years to come that were performed in physiological genetics, even in biochemical genetics, were directed nearly exclusively at the function of genes within specific and well defined environmental conditions. To the extent that the notion of the norm-of-reaction was adopted, it too became a property of the gene or the genotype, rather than that of traits (see, e.g., Clausen, Keck and Hiesey, 1940; 1948).

A significant step towards the reestablishment of the organism as a fundamental and unavoidable entity of life as an evolving continuity was taken by C. H. Waddington in Britain and by I. I. Schmalhausen in the USSR. The work of these investigators should be viewed on the background of *The Modern Synthesis* of the 1940s, that is, the resolution of the conflict between the Mendelian notion of inheritance by discrete, stable, and discontinuous entities and that of Darwinian evolution by incessant, imperceptible small changes of organisms. Although the gene-pool of the Mendelian population was the object of evolutionary change, the individual organism was considered to be the major – actually, the only – vehicle through which Darwinian natural selection was implemented.

Waddington and Schmalhausen emphasized the role of the development of the individual organism as a major constraint on the pattern of evolution of living systems. There was as much intra-organismic interaction that shaped the path of the evolution of a species as there were interactions of organisms with their habitat. However, while Schmalhausen mainly sought to incorporate the development of organisms as a variable that contributes to the Darwinian evolution of the populations' gene pools, Waddington's emphasis was in the opposite direction, namely, on the role of Darwinian evolution in shaping the processes of individual development (see Gilbert, 1994).

Waddington's earlier research in embryology alerted him to the crucial role of epigenetic processes in maintaining stability, yet accommodating changes in circumstances, that made organisms the indispensable discrete links of the evolving chain of life as a continuous phenomenon. Genes provided the long term cumulative memory of the links. Genes, however, would do nothing except where a potential for expressing them was also present. The fundamental mechanism of embryonic development must be one by which the different cytoplasms which characterize the "various regions of the egg, act differentially on the nuclei so as to encourage the activity of certain genes in one region, of other genes in other places" (Waddington, 1957, p. 14). However, whereas Woltereck's attention was directed at the phenotypic response-curves as evidence for the role of the norm-of-reaction in providing the *raw materials* for organismic evolution, Waddington's and Schrnalhausen's primary attention was directed at the norm-of-reaction as providing the *products* of evolution. They stressed the constancy of embryonic development and its regulation as a genetic constraint that had evolved by a process of natural selection. Natural selection worked, within the range of the "habitual environments" of a taxonomic group, for an orderly sequence of developmental processes *in spite* of variation in environmental conditions. Within that range of environmental variations, natural selection buffered developmental responses. "Under the influence of natural selection, development tends to become canalised so that more or less normal organs and tissues are produced even in the face of slight abnormalities of the genotype or of the external environment" (Waddington, 1953, p. 118)

On the one hand, ontogenesis is a process of narrowing the range of possible phenotypes as a function of phylogenetic experience. On the other hand, the norm-of-reaction provides for the phenotypic plasticity that is the potential for extending the range of environments to which the species is adapted: When organisms are exposed to selection pressures that they had not encountered, the interaction of {genotype and new-environment} may respond with specific "acquired characters." These create the opportunity for the phenotypically acquired characters to be tested by natural selection, and – when proved to be "useful" – they may be "assimilated" eventually through replacing genotypically guided processes for the environmentally induced ones and the production of a {new-genotype and environment} interaction.

Waddington endeavored to present a view of life "from above," that overcame the Weismannian dichotomy between the soma and the germ line and delivered the soma from its dead-end destiny. "[With natural living organisms we] find ourselves confronted with … phenomena of historical derivation – … the process of organic evolution; and also with phenomena comparable to those of manufacture – the development of the organism from a fertilized germ to a fully differentiated adult. … It is perhaps the most characteristic feature of biology, and its greatest point of difference from the science of physics and chemistry, that it deals with entities which must be envisaged simultaneously on four different time scales." (Waddington, 1963, p. 25–26)

According to such an approach ontogenesis and phylogenesis should be viewed as two aspects of the same process. The main difference is in their time scale (Falk 1986).

It is not surprising that one of the most outspoken critics of Waddington's ideas was G. C. Williams (1966). Williams's uncompromising campaign for a reductionist adaptationist approach of the New Synthesis denied any attempts to constrain evolutionary thinking beyond that of changes in gene frequencies through selection of its "vehicles," i.e., organisms. He objected to the introduction of developmental constraints as factors that should be considered *per se* in the processes of morphogenesis, since the calculus of gene frequencies sufficed to explain evolution by natural selection of random mutations. Schmalhausen's (1949) and Dobzhansky's (see, e.g., Dobzhansky, 1955) notion of the *adaptive norm* provided a fertile framework for the development of theoretical models of the evolution of polymorphic populations in manifold environments in space and in time. This, however, stifled for many years not only the possibility to consider other units of selection besides the organism, but also the introduction of epigenetic inheritance systems into genetic discourse, i.e., of "[e]verything that leads to the phenotypic expression of the genetic information in an individual" (Jablonka and Lamb, 1995, p. 80, see also pp. 35–36).

Examination of the norm of reaction of specific genotypes under varying experimental conditions (see e.g., Gupta and Lewontin, 1982) highlighted the *unpredictability* of individual phenotypic responses, once the genotype-environment interaction over a wide range was considered (Lewontin, 1974, 1992a). Lewontin especially points out the fallacy inherent in the hopes of analysis of causes through linear models, embodied in the analysis of variance and covariance and path analysis. The analysis of *interacting causes* has been confounded with the concept of discrimination of *alternative causes*. The result of the analysis of variance has a historical (i.e., spatio-temporal) limitation and is not in general a statement about *functional* relations (Lewontin, 1974, p. 403). This analysis is "too specific" in that it completely overlooks the "phenotypic plasticity" exhibited during the life history of each of its individuals, on the other hand it is "too general in that it confounds different causative schemes in the same outcome" (Lewontin, 1974, p. 403). Lewontin (1992a) instead, spells out the route leading from mapping the phenotype space of an individual into a genotype space, to the genotype space of its progeny, and further on, into the progeny's phenotype space, and so on. Thus, he puts the organism squarely back in the evolutionary sequence, rather than viewing it as an evolutionary dead-end epiphenomenon. Lewontin's onslaught on the term "development" as a biasing metaphor of ontogeny summarizes in a nutshell his position: "To describe the life history of an organism as 'development' is to prejudice the entire problematic of the investigation and to guarantee that certain explanations will dominate. 'Development' means literally an unrolling or unfolding … It means the making of an already predetermined pattern immanent in the fertilized egg … All that is required is the appropriate triggering of the process and the provision of a milieu that allows it to unfold. This … reveals the shape of investigation in the field. Genes are everything. The environment is irrelevant except insofar as it allows development." (Lewontin, 1995, p. 261)

It was, however, the challenge of the excessive amount of genotypic variability at the protein and eventually also the DNA level, that forced population geneticists to frantically look for explanations beyond the classic ones of Darwinism and the

Modern Synthesis of genetic polymorphism being either directly adaptive or the price paid for adaptive evolution. Kimura's (1968) and King and Jukes's (1969) proposal that much of the molecular polymorphisms found in natural populations were neutral with respect to natural selection (or beyond its effective reach), stimulated an effort to reexamine the role of adaptive selection in the shaping of individuals and populations.[1] Constraints imposed on genes as members of organized genomes, like those detected by linkage disequilibrium (Lewontin and Kojima, 1960), returned to play a central role in population genetics. Finally the role of the dynamic competition between the need for ontogenetic stability in development and that of phylogenetic malleability in evolution was acknowledged by Eldredge and Gould's (1972) conception of the punctuated equilibrium.

Toward the end of the 1960s there was a feeling among molecular biologists that their spectacular accomplishments may have exhausted the central issues of the life sciences (see, e.g., Stent, 1968). Monod's assertion that what was true for *E. coli* would be true for the elephant (see Judson, 1979, p. 613) reflected the sentiment of many: The organism beyond the cellular level was no meaningful biological entity. Embryogenesis and differentiation did not seem to offer new challenges beyond extending the principles established in the study of bacteria and their phages. However, molecular biology of eukaryotic cells soon encountered many more surprises than could be expected of an elephantiasis-stricken *E. coli*. The role of organisms, and their ontogenesis in biology as an evolving enterprise, demanded a reevaluation of the extreme reductionism of life to molecular survival games of genes as sequences of DNA. The interaction of genetic and epigenetic control of "cellular heredity" in Drosophila, especially as studied for many years by the school of Ernst Hadorn in Zürich, provided the groundwork for molecular thinking of the differentiation of eukaryotic cells. Models of graded (in)activation of genes in differentiation, such as that of the "Bithorax Complex" in shaping the body pattern of the Drosophila embryo (Lewis, 1978), and that of Nüsslein-Volhard (1979) on maternal activation of the ordered differential function of the nuclear genes in the early embryo, could finally lead back to the explication of Weiss's (1940) dictum *"Omnis organisatio ex organisatione"* in molecular terms: Differential *temporal* gene-activation was translated into differential *spatial* organization of the maternal cytoplasm of the oocyte, which in its turn directed the organization of the individual organism's development.

In a sense, the discontinuity of the individual organism, as an independent complex system entity, was challenged. It is, however, exactly this challenge to the organism as an independent entity that emphasized its role in the temporal-to-spatial "recycling" of the organized system. The organism is not a dead-end by-product of germinal continuity, but rather that element which provides the system with the

---

1   The notion of the adaptive "neutrality" of a genetic variant has its roots directly in the genocentric conception, and is antithetic to that of the norm-of-reaction. However, although various authors noted that alleles that were neutral in one context could be non-neutral in another genetic or environmental context, to the best of my knowledge this caveat was not incorporated in the theoretical models of the dynamics of neutral mutations in natural populations.

essential component of Darwinian evolution by selection and mutation, namely, repeated trial and error.

Such a status of the individual multi-cellular organism, as one level in the hierarchy of developmental organization of the evolutionary processes was elaborated by Leo Buss in his *The Evolution of Individuality* (1987). Buss proposed an integrated mechanism of development and evolution of cellular competitive and synergistic interactions: "It is not simply that there is a formal analogy between the selective mechanism of development and the mechanism of natural selection. Rather, it is selection as an evolutionary mechanism that gets frozen into the developmental process by its own success" (Falk and Sarkar, 1992, p. 464). According to Buss the heart of the arguments is "the simple observation that the history of life is the history of the elaboration of new self-replicating entities contained within them (or the incorporation of some self-replicating entities by others). ... At each transition – at each stage in the history of life in which a new self-replicating unit arose – the rules regarding the operation of natural selection changed utterly" (Buss, 1987, p. viii).

The greatest boost to this notion of development as an "ancient interaction between cell-lineages in their quest for increased replication" (Buss, 1987, p. 29) came from an unexpected direction: The reductionist molecular analysis of master genes of developmental pathways in *Drosophila melanogaster*. The highly preserved 183 base-pair DNA sequence in the cluster of homeotic genes of Drosophila, the "homeo-box," was found to be conserved also throughout the animal world, from nematodes to humans (McGinnis et al. 1984a & b). The fact that the homeo-box, and eventually many other sequences of genes that are involved in development, have been highly conserved throughout evolution revolutionized the inter-relation between developmental biology and evolutionary biology. "Early embryos are remarkable for their diversity.... [However, n]o matter how they initiate development, as these embryos establish their body plans and begin to undergo morphogenesis, the conserved Hox rainbows begin to appear." (Kenyon, 1994, p. 176)

In the preface to his book Buss refers to J. T. Bonner's sentiment, who finds it utterly baffling "why one cannot be a reductionist and a holist at the same time" (Buss, 1987, p. vii). Genetics from its inception was considered to be the paradigmatic case of the life sciences of a reductionist science. However, philosophers who tried to formally establish the reductionism of genetics were not very successful (see Schaffner, 1976). On the other hand, experimental scientists, like Woltereck, Goldschmidt, and Waddington, who advocated a holistic approach, were usually marginalized. The undisputed spectacular successes of the modem reductionist experimental *methodology* of molecular biology further entrenched reductionism as a *conception*, to the exclusion of other notions. As late as in 1992 Lewontin had to insist, in a review of a series of books dealing with genetic engineering and the Human Genome Project, that "First, DNA is not self-reproducing, second, it makes nothing, and third, organisms are not determined by it. DNA is a dead molecule, among the most nonreactive, chemically inert molecules in the living world" (Lewontin, 1992b, p. 32). Developmental biology, as an integration of molecular biology and embryogenesis, restores the status of the organism as an entity in the continuum from the holistic view "from above" of the evolution of life, and of the reductionist view "from below" of the functional organization of the molecules of the cells. The convergence of phylogenesis and ontogenesis on the organism is therefore not only an exciting step for experimental biology but also for the epistemology of science.

Literature

Buss, L. W. (1987): The Evolution of Individuality. Princeton: Princeton University Press.

Clausen, J., Keck, D. D., & Hiesey, W. M. (1940): Experimental studies on the nature of species. I. Washington, DC: Carnegie Institute of Washington Publication No. 520.

Clausen, J., Keck, D. D., & Hiesey, W. M. (1948): Experimental Studies on the Nature of Species. III. Washington, DC: Carnegie Institute of Washington Publication No. 581.

Dawkins, R. (1976): The Selfish Gene. Oxford: Oxford University Press.

Dawkins, R. (1982): The Extended Phenotype: The Long Reach of the Gene. Oxford and New York: Oxford University Press.

de Vries, H. (1900): Das Spaltungsgesetz der Bastarde. Berichte der deutschen botanischen Gesellschaft, **18**: 83–90.

Dobzhansky, T. H. (1955): Evolution, Genetics, and Man. New York: John Wiley and Sons.

Eldredge, N., & Gould, S. J. (1972): Punctuated equilibrium: An alternative to phyletic gradualism. In T. J. M. Schopf (Ed.), Models of Paleobiology (pp. 82–115). San Francisco: Freeman, Cooper & Co.

Falk, R. (1986): Can genetics explain development? In E. Ullmann-Margalit (Ed.): The Prism of Science (pp. 165–180). Dordrecht: D. Reidel.

Falk, R. (1995): The struggle of genetics for independence. Journal of the History of Biology, **28**: 219–246.

Falk, R., & Sarkar, S. (1992): Harmony from discord. Biology and Philosophy, **7** (4): 463–472.

Fisher, R. A. (1928): The possible modification of the response of the wild type to recurrent mutation. The American Naturalist, **62**: 115–126.

Gilbert, S. F. (1994): Dobzhansky, Waddington, and Schmalhausen: Embryology and the Modern Synthesis. In M. B. Adams (Ed.), The Evolution of Theodosius Dobzhansky (pp. 143–154). Princeton, NJ: Princeton University Press.

Goldschmidt, R. (1935): Gen und Aussencharakter. III. Biologisches Zentralblat, **55**: 534–554.

Gudding, G. (1996, June): The phenotype/genotype distinction and the disappearance of the body. Journal of the History of Ideas, **57** (3): 525–545.

Gupta A. P., & Lewontin, R. C. (1982): A study of reaction norms in natural populations of *Drosophila pseudoobscura*. Evolution, **36**: 934–948.

Jablonka, E., & Lamb, M. J. (1995): Epigenetic Inheritance and Evolution: The Lamarckian Dimension. Oxford: Oxford University Press.

Johannsen, W. (1909): Elemente der exakten Erblichkeitslehre. Jena: Gustav Fischer.

Johannsen, W. (1911): The genotype conception of heredity. The American Naturalist, **45** (531): 129–159.

Judson, H. F. (1979): The Eighth Day of Creation: Makers of Revolution in Biology. New York: Simon and Schuster.

Kenyon, C. (1994): If birds can fly, why can't we? Homeotic genes and evolution. Cell, **78** (2): 175–180.

Kimura, M. (1968): Evolutionary rate at the molecular level. Nature, **217**: 624–626.

King, J. L., & Jukes, T. H. (1969): Non-Darwinian evolution. Science, **164**: 788–798.

Lewis, E. B. (1978): A gene complex controlling segmentation in Drosophila. Nature, **276**: 565–570.

Lewontin, R. C. (1974): The analysis of variance and the analysis of causes. American Journal of Human Genetics, **26**: 400–411.

Lewontin, R. C. (1992a): Genotype and phenotype. In E. F. Keller & E. A. Lloyd (Eds.), Keywords in Evolutionary Biology (pp. 137–144). Cambridge, MA: Harvard University Press.

Lewontin, R. C. (1992b): The dream of the human genome. New York Review of Books, May 28: 31–40.

Lewontin, R. C. (1995): À la recherche du temps perdu. Configuration, **2**: 257–265.

Lewontin, R. C., & Kojima, K.-I. (1960): The evolutionary dynamics of complex polymorphism. Evolution, **14** (4): 458–472.

Lindsley, D. L., & Grell, E. H. (1968): Genetic Variations of *Drosophila melanogaster*. Washington, DC: Carnegie Institute of Washington Publication No. 627.

McGinnis, W., Garber, R. L., Wirz, J., Kuroiwa, A., & Gehring, W. J. (1984a): A homologous protein-coding sequence in Drosophila homeotic genes and its conservation in other metazoans. Cell, **37**: 403–408.

McGinnis, W., Levine, M.S., Hafen, E., Kuroiwa, A., & Gehring, W. .J (1984b): A conserved DNA sequence in homoeotic genes of the Drosophila Antennapedia and bithorax complexes. Nature, **308** (5958): 428–433.

Muller, H. J. (1950): Evidence of the precision of genetic adaptation. The Harvey Lectures, **43**: 165–229.

Nüsslein-Volhard, C. (1979): Maternal effect mutations that alter the spatial coordinates of the embryo of *Drosophila melanogaster*. In S. Subtelny & I. Konigsberg (Eds.), Determinants of Spatial Organization (pp. 185–211). New York: Academic Press

Rieger R., Michaelis, A., & Green, M. M. (1976): Glossary of Genetics and Cytogenetics (4th ed.). Berlin: Springer-Verlag.

Roll-Hansen, N. (1978): The genotype theory of Wilhelm Johannsen and its relation to plant breeding and the study of evolution. Centaurus, **22**: 201–235.

Sapp, J. (1987): Beyond the Gene. Cytoplasmic Inheritance and the Struggle for Authority in Genetics. New York: Oxford University Press.

Schaffner, K. (1976): Reduction in biology: Prospects and problems. In R. S. Cohen, C. A. Hooker, G. Pearce, A. C. Michalos, & J. W. van Evra (Eds.), Boston Studies in the Philosophy of Science: 32. Proceedings of the 1974 Biennial Meeting of the Philosophy of Science Association (pp. 613–632). Dordrecht: Reidel.

Schmalhausen, I. I. (1986 [1949]): Factors of Evolution. The Theory of Stabilizing Selection. Chicago and London: University of Chicago Press.

Stent, G. S. (1968): That was the molecular biology that was. Science, 160: 390–395.

Timoféeff-Ressovsky, H. A., & Timoféeff-Ressovsky, N. W. (1926): Über das phaenotypische Manifestieren des Genotyps. II Über idio-somatische Variationsgruppen bei *Drosophila funebris*. Wilhelm Roux's Archives of Developmental Biology, **108**: 146–170.

Vogt, O. (1926): Psychiatrisch wichtige Tatsachen der zoologisch-botanischen Systematik. Zeitschrift für die gesamte Neurologie und Psychiatrie, **101**: 805–832.

Waddington, C. H. (1953): Genetic assimilation of an acquired character. Evolution, **7**: 118–126.

Waddington, C. H. (1957): The Strategy of the Genes. A Discussion of Some Aspects of Theoretical Biology. London: Unwin Books.

Waddington, C. H. (1963): The Nature of Life. London: Unwin Books.

Weiss, P. (1940): The problem of cell individuality in development. The American Naturalist, **74**: 34–46.

Williams, G. C. (1966): Adaptation and Natural Selection. Princeton, NJ: Princeton University Press.

Woltereck, R. (1909): Weitere experimentelle Untersuchungen über Artveränderung, speziell über des Wesen quantitativer Artunterschiede bei Daphnien. Verhandlungen der Deutschen Zoologischen Gesellschaft, **19**: 110–173.

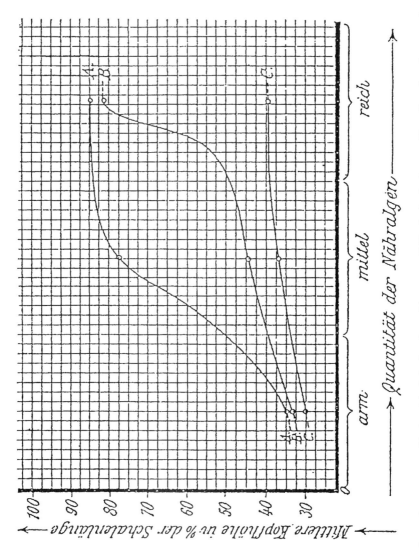

Figure 1. Norm of reaction curves of helmet-height for three biotypes of *Hyalodaphnia cucullata* females, in poor, intermediate and rich quantities of nutrients. (after Woltereck, 1909).

The organism's place in evolution: Darwin's views and contemporary organismic theories

Franz M. Wuketits

1. Introduction

Darwin's theory is still to be regarded as the best starting point in evolutionary thinking. Its hard core – and the hard core of any Darwinian tradition – is the theory of natural selection, and it is to be expected that Darwinian theories "will always include natural selection as a fundamental element" (Gayon, 1994, p. 106). However, during the last 25 years or so the natural-selection conceptions inherent in Darwin's views and, particularly, the views of Neo-Darwinists and the advocates of synthetic theory have been criticized by many evolutionists. New theories emerged, and even extremist views like 'evolution without selection' (Lima-de-Faria, 1990) have been formulated and defended. The new theories can be regarded as nonadaptationist – sometimes indeed antiadaptationist – research programs that go beyond the well-established synthetic theory that has long had the status of a textbook theory of organic evolution. Their advocates put the organism in the center of evolutionary thinking and consider organismic constraints as important to any evolutionary change; the organism itself has thus become a major factor of evolution.

In this situation it seems quite interesting to go back once again to Darwin and to notice that he himself was not completely satisfied with his solution of some problems of evolution, and that parts of his work have been somehow distorted by the ardent advocates of adaptationism inherent in the hard-core version of synthetic theory or 'the Modern Synthesis' (Huxley, 1942). Darwin himself was not a mere selectionist and adaptationist!

In this paper I shall give, first, a very brief review of what organismic theories of evolution today actually mean, and, especially, how their proponents evaluate the organism's role in evolutionary processes; second, I shall speculate on how Darwin himself would meet such theories and how he would cope with the notion of organismic constraints. Sure, questions like 'Would Darwin, if he lived today, become an advocate of any of the organismic theories?' are unwarranted for reasons that are well known in historical science. But we have to look carefully at Darwin's work – which does offer some evidence that he was quite familiar with 'organismic thinking' in a wider sense.

2. Organismic Theories of Evolution

2.1 General reflections

Organismic theories of evolution are generally stressing the role of the organism in evolutionary change and are based on the assumption that the living being as a complex, hierarchically organized system acts, as it were, as an important factor in evolution. A key notion is internal selection (see e.g. Wuketits, 1985b, for discussion). It was already pointed out by White in his "Internal Factors in Evolution". "I believe", he wrote, "that the principle of internal selection is important and timely because it is the expression of the contemporary experimental and theoretical concern with structure. If that is so, then those that apply their energies to clarifying and applying the principle are likely to be rewarded by discoveries relating to the coordinative conditions which characterize living systems." (White 1965, p. 78). He assumed that "many properties, especially those of coordination, may be the result of internal selection" (White 1965, p. 32). However, he did not disregard external, Darwinian selection, but maintained that an internal or intraorganismic selective force is the additional specific factor in evolutionary change.

Years later, Lewontin summarized the interplay of external (environmental) and internal (intraorganismic) factors as follows: "Organisms … both make and are made by their environment in the course of phylogenetic change, just as organisms are both the causes and consequences of their own ontogenetic development. The alienation of internal and external causes from each other and of both from the organism, seen simply as passive result, does not stand up under even the most casual survey of our knowledge of development and natural history." (Lewontin 1983, p. 284).

Together with Gould, Lewontin had pointed already earlier to the failure of the adaptationist program to conceive of the evolutionary process as a process with passive organisms instead of seeing living systems as integrated wholes with specific constraints delimiting pathways of (evolutionary) change and thus being more interesting than environmental selective forces (Gould and Lewontin, 1978). Moreover, he and Gould had laid particular emphasis on developmental, ontogenetic constraints that have meanwhile received much attention by many researchers and have even been considered as elements of a new evolutionary synthesis (Müller, 1994).

It is apparent that any organismic theory of evolution leads to the assertion that evolution is not a linear step-by-step process with an accumulation of adapted traits in the organisms, but rather a process of 'self-organization' with self-regulation as a basic principle in the life of any living system. Adaptation may play its role – although, as we shall see, the proponents of one of the organismic theories have completely abandoned the adaptationist program –, however adaptability is not and cannot be defined by the organisms' environment, but by the organisms themselves (Wagner, 1985). In other words: The extent of evolutionary changes, their particular direction, and the degree of adaptation as a result of these changes are largely determined by the organization of the living beings in question. Hence, selection cannot be seen as an agenda for 'environmentalism' or, as Gould puts it: "Selection may supply all immediate direction, but if highly constraining channels are built of non-adaptations, and if evolutionary versatility resides primarily in the nature and extent of nonadaptive pools, then 'internal' factors of organic design are an equal partner with selection." (Gould 1982, p. 384).

However, such internal factors do not necessarily contradict natural selection in a strict Darwinian sense. They, too, can be regarded as selective forces, forces acting inside the organism where all the organs mutually represent 'environment' (Wagner, 1985).

In what follows I give a short review of three contemporary organismic theories of evolution: systems theory, critical theory, and the theory of punctuated equilibria. There are other theories deserving the label 'organismic', but, for brevity's sake, will not be considered, as for example 'epigenetic theory' (Lovtrup, 1982) which supports 'saltationism', i.e. the position that macroevolutionary phenomena cannot be properly explained by Darwinian selection and are mainly to be attributed to 'saltations' (see e.g. Schindewolf, 1950). (For more details and critical comments see also Wuketits, 1988.)

2.2 Systems theory

The systems theory of evolution (see, e.g., Riedl, 1975, 1977; Wagner, 1983, 1985, 1986; Wuketits, 1985a, 1987, 1988) takes Darwinian selection seriously. However, its advocates maintain that this type of selection does not suffice to explain many evolutionary phenomena, particularly at the level of macroevolution. The central elements of the theory can be summarized as follows:

(1) Organisms as active systems are not simply molded by their environment; they are not marionettes hanging on the strings of environmental selection, but able to determine, to a certain extent, their own surroundings. In fact, there is a constant interaction between any living being and its outer world.

(2) Organisms are hierarchically organized, multilevel systems, and their levels and parts are mutually related and linked together by feedback loops and regulatory principles. Not only do the parts determine the whole organism, but the organism vice versa determines and constrains the structure and function of its parts, so that there is a constant flux of cause and effect in both directions – up and down the levels of the hierarchically organized system.

(3) Evolution results from internal and external (intraorganismic and environmental) selection that do not work independently but together build the systems conditions of evolutionary change. Evolution is to be considered as a complex process of dynamical 'self-organization'. As self-regulative systems, organisms themselves strongly influence the paths of their own evolution, like a river channeling its own river-bed (Wagner, 1985).

(4) There are complex interactions between the genotype and phenotypic variation. It can be assumed that species-specific genetic traits are modifiable to such an extent that a variety of phenotypic characters becomes possible. Some characters may be preserved as fundamental traits of the respective "Bauplan", others may vary to a significant degree. Thus, there is no fundamental discontinuity in evolution, and microevolution and macroevolution are linked together.

(5) The systems-theoretic approach is nonreductionist and nonatomist, basing on the old wisdom that the whole is more than the sum its parts. However, an important empirical basis for the systems theory of evolution is population genetics.

2.3 Critical theory

Critical theory is the name of the evolutionary conceptions launched by the so-called 'Frankfurt School' (see, e.g., Gutmann, 1972, 1979, 1989, 1994; Gutmann and Bonik, 1981; Gutmann and Edlinger, 1994; Gutmann and Weingarten, 1989). The representatives of this school have abandoned the notion of adaptation and reject any effect of the environment upon the organism. While the systems theory of evolution is an extension of Darwin's theory (and synthetic theory), the critical theory contradicts any Darwinian approach to understanding evolution. Its central tenets are the following:

(1) Organisms are hydraulic, energy processing systems or, in other words, mechanically coherent constructions built according to the principles of hydraulics. They are not adapted to their environment(s), but are self-maintaining, self-acting systems that spontaneously 'step in' the energy flow of their environment(s) and take whatever they need to feed their own machinery.

(2) Organic evolution is to be seen as determined by functional constraints, i.e. internal functional (and constructional) conditions. The environment does not mold any organism, living systems are rather 'looking' for their appropriate environments. The living organism is largely autonomous and thus independent of external influences. (This reminds us of 'constructivism' as it has been defended for example by Maturana and Varela [1980]. In fact, the critical theory of evolution is a constructivist approach and therefore sometimes called 'constructional theory'.)

(3) Evolution is an active process. This tenet is in accordance with a similar claim by the advocates of systems theory, however, what divides the two groups of theorists is the assumption (of the defenders of critical theory) that processes of evolutionary transformations and environmental changes in earth history have to be reconstructed as parallel but actually decoupled phenomena. Hence, fossils do not play a major role in critical theory but are left as mere time marks in biostratigraphy.

(4) What the advocates of critical theory are practically doing is producing nonhistorical models of historical processes.

2.4 Theory of punctuated equilibria

The advocates of this theory (e.g., Benton, 1990; Eldredge and Cracraft, 1980; Eldredge and Gould, 1972; Gould, 1989; Stanley, 1980) are usually paleontologists using the fossil record against phyletic gradualism which is deeply rooted in nineteenth-century evolutionism and supported by old and venerable philosophical conceptions of nature (natura non facit saltus). According to the gradualist view, evolution is a more or less linear process, a step-by-step process of ever-better adaptations. Characterizing the very meaning of punctuated equilibrium, Gould (1984, p. 260) writes: "We proposed the theory of punctuated equilibrium largely to provide a different explanation for pervasive trends in the fossil record. Trends, we argued, cannot be attributed to gradual transformation within lineages, but must arise from the different success of certain kinds of species. A trend, we argued, is more like climbing a flight of stairs (punctuation and stasis) than rolling up an inclined plane."

In particular, the claims inherent in the theory of punctuated equilibrium are basically the following:

(1) Paleontologists are often confronted with tachytely, fast-rate phylogenetic development or the abundant appearance of new species within a particular geological period. This phenomenon contradicts the gradualist view. One has to expect that phylogenetic changes occur 'in batches' from time to time, that the equilibrium is interrupted or punctuated. But not only do species from time to time appear at a great pace, the disappearance or extinction sometimes also comes abruptly.

(2) Darwinian explanations in a strict sense do not suffice to explain these phenomena and especially the phenomena of macroevolutionary changes (including extinction). Adaptationism does not offer the proper explanation of evolutionary transformations for these transformations concern complex living systems with their most specific architecture and developmental (ontogenetic) constraints (Gould and Lewontin, 1978).

(3) Studying evolution and reconstructing the paths of phylogenetic development means to carefully consider whole organisms – and not just particles and genes. Similar to the systems-theoretic approach, the theory of punctuated equilibria is nonreductionist and nonatomist.

(4) Macroevolution deserves particular attention. In a way, it can be studied separately, as a phenomenon to a certain degree decoupled from microevolutionary processes.

3. Darwin's Organismic Views of Evolution

As I stated at the outset, this part of the present paper is somehow speculative, however, if one reads Darwin's work carefully, one will easily find ideas that point to an organismic conception of evolution and the suggestion that Darwin, after all, would embrace an organismic theory.

3.1 Darwin's theory revisited

Darwin's theory has been frequently discussed by evolutionists and historians and philosophers of science. More recently, Ernst Mayr, one of the architects of synthetic theory, published a book on Darwin's arguments which also includes a (sometimes harsh) reaction to the critics of Darwinian theories who have obviously irritated the old master (Mayr, 1991). It is not – and cannot be – the task of this paper to pick up once again all the points of controversy that have arisen in endless discussions and debates ever since the first edition of "The Origin of Species" in 1859 and to deal with all the misconceptions and misunderstandings with which 'the Darwinists' had and still have to cope.

What seems undisputed is that Darwin's theory of evolution and its causal explanation includes at least the following elements:

– The idea of struggle for life caused by limited resources and a tendency towards exponential reproduction.

– The idea of the survival of the fittest based on the observation that the individuals of any species vary to a significant extent.

– The assumption that natural selection is the essential motor of evolutionary change.

– The statement that natural selection acts upon different stages in the life of any organism and thus "can modify the egg, seed, or young, as easily as the adult" (Darwin, 1859 [1958, p. 128]).

– The principle of continuity according to which evolution occurs by gradation, step by step, so to speak, without great or sudden modifications.

– The conjecture that natural selection adapts and improves some varieties of each species.

These elements can also be found in more recent Darwinian theories, particularly in the synthetic theory of evolution.

However, it is evident that Darwin's original theory has undergone some revisions and modifications with regard to many of its details – especially in the sphere of genetics –, so that Darwinism itself has evolved (see, e.g. Mayr, 1991; Stebbins and Ayala, 1981; Wieser, 1994; Wuketits, 1995). For this reason it is sometimes difficult to differentiate between 'the model' (Darwin's original theory) and 'the copy' (Darwinism) (see Gayon, 1994). The synthetic theory which has become predominant in evolutionary thinking after the 1940s has veiled some of Darwin's original ideas.

3.2 Darwin's organismic views

I seriously call into doubt the wide-spread belief that Darwin was a strict selectionist and adaptationist, no matter that he indeed considered selection as the essential driving force in evolution. Obviously, he could not sufficiently explain all phenomena of life via natural selection. An example are rudimentary organs which compelled him to resort to Lamarck's doctrine of inheritance. Thus, in a short paper on rudimentary structures in cirripedes (1873), Darwin came to the following conclusion: "With respect to the means by which so many of the most important organs in numerous animals and plants have been greatly reduced in size and rendered rudimentary, or have been quite obliterated, we may attribute much to the inherited effects of the disuse of parts." (Darwin, 1977, vol. II, p. 180)

But already in "The Origin of Species" Darwin had speculated on this problem and stated the following: "If, for instance, it could be proved that every part of the organisation tends to vary in a greater degree towards diminution than towards augmentation of size, then we should be able to understand how an organ which has become useless would be rendered, independently of the effects of disuse, rudimentary and would at last be wholly suppressed; for the variations towards diminished size would no longer be checked by natural selection." (Darwin, 1958, p. 423)

Moreover, we read in the same passage: "The principle of the economy of growth ... by which the materials forming any part, if not useless to the possessor, are saved as far as possible, will perhaps come into play in rendering a useless part rudimentary. But this principle will most necessarily be confined to the earlier stages of the process of reduction." (Darwin, 1958, p. 423)

It is interesting to see that Darwin was aware of phenomena which are not necessarily 'checked by natural selection'. I do think that he would have used the concept of intraorganismic constraint if it had been known to him.

Another example of Darwin's possible reflections on what we now call 'organismic constraints' is his speculation on the phylogenetic development of the eyes. Certainly, Darwin "urged in the Origin that natural selection accomplished most of the work done in evolution." (Richards, 1992, p. 84), but with regard to such complex phenomena like eyes he was not completely satisfied with his theory. At least he admitted that one could doubt the power of natural selection as an overall explanation of everything and anything in the living world. "The belief", he remarked, "that an organ so perfect as the eye could have been formed by natural selection, is enough to stagger any one." (Darwin, 1859 [1958, p. 18]). Although Darwin griped this belief on principle grounds, he did not miss to see and understand the meaning of complexity and Bauplan. I cannot go into details here but should like to suggest that Darwin, particularly regarding complex phenomena like the eye, was generally prepared to look for mechanisms other than natural selection, even though he would of course never have diminished the importance of selective forces.

I agree to Gould and Lewontin (1978) that Darwin's own approach to identifying mechanisms of evolutionary change was pluralistic. Just remember the important role that variation played in his view of evolution. After all, in the last edition of "The Origin of Species" Darwin noted that he had never attributed the modification of species exclusively to natural selection. This was only the wrong impression of many of his critics. Wrong is also the impression that Darwin made external conditions responsible for variation. He explicitly stated: "Naturalists continually refer to external conditions, such as climate, food, etc., as the only possible source of variation. In one limited sense ... this may be true; but it is preposterous to attribute to mere external conditions, the structure, for instance, of the woodpecker, with its feet, tail, beak, and tongue, so admirably adapted to catch insects under the bark of trees." (Darwin, 1859 [1958, p. 28]).

What else, if not intraorganismic constraints, could he have had in mind when he wrote this? The concept of internal selection, as it is used today especially by the advocates of the systems theory of evolution would, I suppose, be quite in line with Darwin's intentions. Darwin did not underestimate the role of the organism in evolution, and it would be a fundamental misunderstanding of his work as a whole, would one appreciate him merely for his ideas on natural selection (as environmental selection) and adaptation.

One should, by the way, also keep in mind that the early Neo-Darwinists, particularly August Weismann – the very founder of Neo-Darwinism – were prepared to take up internal selection as an important aspect of the organization and evolution of living systems. Weismann (1904) explicitly spoke of 'intraselection' (Intraselektion) to characterize the differences in the development of organs in any individual. And Wilhelm Roux, the pioneer of the physiological study of ontogenetic development, refered to self-regulation as a characteristic capacity of all organisms and applied the concept of selection to the inner organization of living systems (Roux, 1914).

Thus we can state that organismic theories have had a long and venerable tradition within the Darwinian view. In fact, if we re-read Darwin and his ideas we are left with the impression that many of his successors have delivered only parts of the massive corpus of his ideas. Let's continue studying the very message of Darwin and applying his view to analyzing the complex phenomena of organismic evolution.

Literature

Benton, M. J. (1990): Evolutionsforschung aus der Sicht des Paläontologen. In: U. Jüdes, G. Eulefeld, T. Kapune (eds.), Evolution der Biosphäre, 9–49. Hirzel, Stuttgart.

Darwin, Ch. (1859/1958): On the Origin of Species by Means of Natural Selection. Murray, London/New American Library, New York, Toronto.

Darwin, Ch. (1977): The Collected Papers of Charles Darwin. 2 vols. (P. H Barrett, ed.). The University of Chicago Press, Chicago, London.

Eldredge, N., Cracraft, J. (1980): Phylogenetic Patterns and the Evolutionary Process. Methods and Theory in Comparative Biology. Columbia University Press, New York.

Eldredge, N., Gould, S. J. (1972): Punctuated Equilibria: An Alternative to Phyletic Gradualism. In: T. J. M. Schopf (ed.), Models in Paleobiology, 82–115. Freeman, San Francisco.

Gayon, J. (1994): What does "Darwinism" Mean? Ludus Vitalis 2 (2), 105–118.

Gould, S. J. (1982): Darwinism and the Expansion of Evolutionary Theory. Science 216, 380–387.

Gould, S. J. (1984): Hen's Teeth and Horse's Toes. Further Reflections in Natural History. Penguin Books, Harmondsworth.

Gould, S. J. (1989): Punctuated Equilibrium in Fact and Theory. J. Social Biol. Struct. 12, 117–136.

Gould, S. J., Lewontin, R. C. (1978): The Spandrels of San Marco and the Panglossian Paradigm: A Critique of the Adaptationist Programme. Proc. Royal Soc., Series B, 205, 581–598.

Gutmann, W. F. (1972): Die Hydroskelett-Theorie. Aufs. u. Reden d. Senckenberg. naturforsch. Ges. 21, 1–91.

Gutmann, W. F. (1979): Entwickelt sich ein neues Evolutionsverständnis? Das Analogie-Denken Darwins und die physikalistische Evolutionstheorie. Biol. Rdsch. 17, 84–99.

Gutmann, W. F. (1989): Die Evolution hydraulischer Konstruktionen. Organismische Wandlung statt altdarwinistischer Anpassung. Kramer, Frankfurt a. M.

Gutmann, W. F. (1994): Evolution von Konstruktionen. Die Frankfurter Theorie. In: W. F. Gutmann, D. Mollenhauer, D. S. Peters (eds.), Morphologie & Evolution, 317–337. Kramer, Frankfurt a. M.

Gutmann, W. F., Bonik, K. (1981): Kritische Evolutionstheorie. Ein Beitrag zur Überwindung altdarwinistischer Dogmen. Gerstenberg, Hildesheim.

Gutmann, W. F., Edlinger, K. (1994): Neues Evolutionsdenken: Die Abkopplung der Lebensentwicklung von der Erdgeschichte. Archaeopteryx 12, 1–14.

Gutmann, W. F., Weingarten, M. (1989): Das Problem der Form biologischer Systeme. In: K. Edlinger (ed.), Form und Funktion. Ihre stammesgeschichtlichen Grundlagen, 21–34. WUV-Universitätsverlag, Wien.

Huxley, J. (1942): Evolution. The Modern Synthesis. Allen & Unwin, London.

Lewontin, R. C. (1983): Gene, Organism and Environment. In: D. S. Bendall (ed.), Evolution from Molecules to Men, 273–285. Cambridge University Press, Cambridge, London, New York.

Lima-de-Faria, A. (1990): Evolution ohne Selektion: Form und Funktion durch Autoevolution In: U. Jüdes, G. Eulefeld, T. Kapune (eds.), Evolution der Biosphäre, 105–122. Hirzel, Stuttgart.

Lovtrup, S. (1982): The Epigenetic Theory. Rivista Biol. **75**, 231–255.

Maturana, H., Varela, F. (1980): Autopoesis and Cognition. The Realization of the Living. Reidel, Dordrecht, Boston, London.

Mayr, E. (1991): One Long Argument. Harvard University Press, Cambridge, Mass.

Müller, G. B. (1994): Evolutionäre Entwicklungsbiologie: Grundlagen einer neuen Synthese. In: W. Wieser (ed.), Die Evolution der Evolutionstheorie. Von Darwin zur DNA, 155–193. Spektrum Akademischer Verlag, Heidelberg, Berlin, Oxford.

Richards, R. J. (1992): The Meaning of Evolution. The Morphological Construction and Ideological Reconstruction of Darwin's Theory. The University of Chicago Press, Chicago, London.

Riedl. R. 1975): Die Ordnung des Lebendigen. Systembedingungen der Evolution Parey, Berlin, Hamburg.

Riedl, R. (1977): A Systems-Analytical Approach to Macro-Evolutionary Phenomena. Quart. Rev. Biol. **52**, 351–370.

Roux, W. (1914): Die Selbstregulation. Ein charakteristisches und nicht notwendig vitalistisches Vermögen aller Lebewesen. Nova Acta Leopoldina **100** (2), 1–91.

Schindewolf, O. H. (1950): Grundfragen der Paläontologie. Geologische Zeitmessung, organische Stammesentwicklung, biologische Systematik. Schweizerbart, Stuttgart.

Stanley, S. M. (1980): Macroevolution. Pattern and Process. Freeman, San Francisco.

Stebbins, G. L., Ayala, F. J. (1981): Is a New Evolutionary Synthesis Necessary? Science **213**, 967–971.

Wagner, G. P. (1983): On the Necessity of a Systems Theory of Evolution and Its Population Biologic Foundation. Acta Biotheor. **32**, 223–226.

Wagner, G. P. (1985): Über die populationsgenetischen Grundlagen einer Systemtheorie der Evolution. In: J. A. Ott, G. P. Wagner, F. M. Wuketits (eds.), Evolution, Ordnung und Erkenntnis, 97–111. Parey, Berlin, Hamburg.

Wagner, G. P. (1986): The Systems Approach: An Interface Between Development and Population Genetic Aspects of Evolution. In: D. M. Raup, D. Jablonski (eds.), Patterns and Processes in the History of Life, 149–165. Springer, Berlin, Heidelberg.

Weismann, A. (1904): Vorträge über Deszendenztheorie. Fischer, Jena.

Whyte, L. L. (1965): Internal Factors in Evolution. Tavistock Publications, London.

Wieser, W. (1994): Gentheorien und Systemtheorien: Wege und Wandlungen der Evolutionstheorie im 20. Jahrhundert. In: W. Wieser (ed.), Die Evolution der Evolutionstheorie. Von Darwin zur DNA, 15–48. Spektrum Akademischer Verlag, Heidelberg, Berlin, Oxford.

Wuketits, F. M. (1985a): Die systemtheoretische Innovation der Evolutionslehre. Historische und erkenntnislogische Voraussetzungen einer Theorie der Systembedingungen der Evolution. In: J. A. Ott, G. P. Wagner, F. M. Wuketits (eds.), Evolution, Ordnung und Erkenntnis, 69–81. Parey, Berlin, Hamburg.

Wuketits, F. M. (1985b): Zum Konzept der "inneren" Selektion. Stellungnahme zu einer evolutionstheoretischen Kontroverse. Paläont. Z. **59**, 35–41.

Wuketits, F. M. (1987): Evolution als Systemprozeß: Die Systemtheorie der Evolution. In: R. Siewing (ed.), Evolution. Bedingungen – Resultate – Konsequenzen, 453–474. Fischer, Stuttgart.

Wuketits, F. M (1988): Evolutionstheorien. Historische Voraussetzungen, Positionen, Kritik. Wissenschaftliche Buchgesellschaft, Darmstadt.

Wuketits, F. M. (1995): Darwin und die Selektionstheorie: Gestern, heute – und morgen? In: J. Cimutta, F. M. Wuketits (eds.), Lebt Darwins Erbe? Fragen und Standpunkte zur Evolutionstheorie, 25–54. A. Lenz Verlag, Neustadt i. Rbge.

The development of organismic structure and the philosophy behind

CHRISTIAN KUMMER

Introduction

Modern Biology tries to understand life and organisms by means of mechanistic interpretation. In times of Descartes, it was the whole living being, which was thought to be a machine and, consequently, should function without any kind of "soul" (an animal without "anima"), which since the days of Aristotle was regarded as the characteristic of life. Nowadays, the machine theory of life ranges on a molecular level, and thus suggests organisms to be the highly complex outcomes ("systems") of intracellular interaction networks. In opposition to this commonly held view (and somehow like an alien element among the "orthodox" believers in the dominant biological doctrine), there is the "Frankfurt Theory" of constructive morphology which disputes the sufficiency of molecular and genetical determination of organization, and establishes a holistic view of what is the constitutive of an organism. It understands the organism as a machine, too, (or more precisely, as a "hydraulic" one), but looks for the constructive principles which constitute such hydraulic and energy transforming "machines".

Since it can hardly be seen how organisms work by principles in order to achieve their proper form, there remains a deep epistemological gap between the explanatory claims of constructive morphology and molecular biology. To mediate these two positions, a short glance at the history of philosophy may be helpful, because this will present us with an alternative organismic theory, namely that of Aristotle, which conceives the principles of organismic constitution as the real (albeit metaphysical) causes of animal development. This might help to clarify the ontological impact or deficiency, respectively, inherent in both the molecular and the constructive approach of explanation.

As it intends to elucidate the rearrangement of organizations, constructive morphology deals with both development and evolution. It is the same hydraulic principle which permits the morphogenetic movements during ontogeny, and constitutes the constructive transitions from one phylogenetic type to another. To understand not only the rules but the causes that govern the formative process, we shall concentrate on developmental biology, because this is the field where Aristotle's theory of becoming (*kinesis*) still seems to be valid, and, of course, he did not think in terms of evolution.

1. The mechanistic theory of becoming

We should bear in mind that Aristotle has worked out his theory in a situation very similar to the intention of constructive morphology: both rose in explicit opposition to the appeal of a mechanistic world view.

The great problem of early Greek philosophy was how to understand becoming, i.e. the changing of things in our empirical world. We may find it ridiculous to question such an every-day's experience. Such an attempt to undermine the world of common sense seems to be inseparably linked to an odd kind of essentialistic thinking. But from a logical point, we may soon find that it is not so easy to explain why we are convinced of the identity or difference, respectively, of "objects" underlying our sensual perception. If we want to distinguish the "real" from "seeming", or better, objective from subjective states of change, we need a fixed point of reference.

The Archimedean point of epistemology suggested by the Greek atomists, as Leukippus and, above all, Democritus, is the eternity of matter. To allow material things to undergo mutation they think things to consist of last indivisible and unchanching units, the so-called atoms, and assume empty space and (perpetual) motion to enable the atoms to aggregate in different compositions. In that way it seems possible to reduce all qualitative change to a mere quantitative effect of replacement of particles.

Though the question of mutability is no longer a theoretical problem to modern science which has become accustomed to conceive everything within the framework of the evolutionary paradigm, it still adheres to the same explanatory ideal of the old Greek atomists:

Increasing complexity is to be explained fully in terms of the motion of underlying particles.

Some terms of this sentence need further explication.

Increasing complexity denotes a structure becoming more complicated. Obviously, there are structural changes, as degradation or decay, where the knowledge of the specific cause normally is not of particular interest. What needs a proper explanation is a system changing to a *higher* degree of organization or order, for both physical and philosophical (or, let us say, metaphysical) reasons. From the viewpoint of thermodynamics, every system counteracting the gradient of entropy requires an additional supply of energy to do so. Philosophically speaking, performing a higher level of organization indicates a higher state of being, which cannot emerge from nothing, and so calls for an adequate cause.

Complexity characterizes the connections between a large number of components of a 'system'. According to Hans Driesch[1] one can distinguish 4 different meanings of complexity on account of its causation:

• complexity as *cumulation*, which comprises all modern forms of material self-organization;

---

1   *Philosophie des Organischen*, Leipzig (1928), 375. – We do not agree with the reasoning of neovitalism, but sometimes we find the conceptual distinctions of Hans Driesch very helpful.

- complexity by *preformation*, which characterizes the change of form as a mere transition from an invisible to a visible state;
- complexity by *information*[2], when the acquisition of a new structure results from the realization of an *intentional* state of form (= pre-existing program), which is the classical view of genetic determinism;
- complexity as *organization*, where a proper *formal cause* is introduced in order to explain the quality of structure. However, in contrast to Driesch this formal cause must not be understood as a special factor within the system, but rather as the underlying unity expressed by the internal and external relations of the form-giving elements. This concept is due to Aristotle, and transcends the realm of mechanistic interpretation.

Particles are specified by the lowest level of explanation characteristic of a system. Thus, Dubois-Reymond thought the knowledge of the actual position of every atom within a totally determined universe would enable the demon of Laplace to predict every future event[3]. As the processes of life are performed by specific molecules, development of organic structure is fully explained by dissecting it completely into its molecular interdependencies. In this way increasing complexity is conceived as a sequence of integrative levels, and explanation is reductive in the sense that the elementary level is assumed to cause the characteristic properties of a system.

Motion of particles implies, then, the complete reduction of changing qualities to that motion. In Greek, *kinesis* (motion) includes a *qualitative* connotation, which means changing something in the sense of a whole thing becoming something different. Using this term, Democritus could insinuate an acceptable explanation of qualitative change without presenting a sufficient principle. According to him, new properties of things are brought about by new states of accumulation of particles, and these accumulations are caused by random collisions of those particles, and need no additional formal cause. And it remains up to the present an explanatory ideal of mechanistic thought to reduce transformation to complete random transposition, regardless of various modern attempts to introduce some kind of formal causes which all are, deliberately or not, based on conceptual equivocations similar to those involved in the term of motion. So we use the expression 'movement' to characterize the mechanistic understanding of causality in its strictest sense.

---

2 Note that our terminology is not in all respects the same Driesch was using. What we call "complexity by preformation" in Driesch's own terms is *Scheinentwicklung*; "complextiy by information" is equivalent to Driesch's *maschineller* or *präformierter Entwicklung*; and "complexity as organization" corresponds to some degree to *entelechiale* or *nicht-präformierte Entwicklung*.

3 Über die Grenzen des Naturerkennens (1872), in: E. Du Bois-Reymond, *Vorträge über Philosophie und Gesellschaft.* Hamburg (1974), 59.

2. Aristotle's holistic concept of substantial change

As complexity signifies not only the complication of matter caused by an increasing amount of elements, but also the emergence of well-ordered structure, both definite and reproducible, pure movement of elements, without any leading "idea", provides no sufficient cause for reproducible formative effects. To overcome this deficiency inherent mechanism Aristotle does not start with particles which then acquire systemic coherence, but with the *whole* visible thing undergoing a change, and calls such a concrete entity a *substance*, Greek *ousia*. Usually, this substance concept is regarded as completely static, since it defines the species of things as eternal and given by their conceptual essence[4].

However, Aristotle's substance concept contains more than this essentialistic element. It also involves a process of material realization of concrete forms that furnishes natural things (*physei onta*) with the potential of self-organization[5]. To refer to this dynamic aspect of accomplishing the proper form of a concrete natural thing, Aristotle uses the expression "nature" (*physis*). It is the nature of a substance (and, especially, of a living being) that establishes its proper form by controlling the elementary processes.

Since the nature of a concrete thing is nothing else but the form of the substance in terms of its self-determination, it is possible to claim with Aristotle that it is form which causes form. In order to give this apparently tautological statement a more concrete sense we point out a twofold meaning that is attached to *ousia*. In the first place it signifies the concrete being with the actual form it has; secondly it stands for the perfect state of form this being could have[6]. This formal perfection is the "idea" (*eidos*) of a substance, and the dynamic of nature results from the actual difference between this ideal state of perfection and its concrete realization, performing what we could call a kind of "morphogenetic field" (albeit in an idealistic sense), allowing the elements to assemble in a specific order.

---

4   The examples presented by Aristotle in Met. VII 8 may justify this misinterpretation: it is not the craftsman who generates a sphere, since the form, or let us better say, the geometry of the sphere precedes all its material realizations (1033a 28). Similarly, man is always generated by man (1033b 31), what proves the species of man to be eternal, regardless of the limited lifespan of human individuals.

5   It is one of the merits of R.L. Fetz's book: *Whitehead: Prozeßdenken und Substanzmetaphysik*, Freiburg/München 1981, to point out this dynamic aspect of the Aristotelean substance. What we have called "self-organization", referring to a modern expression, in the author's own words is described as "Selbstbestimmung" (self-determination), *op. cit.*, p. 213. – For further distinction between different kinds of self-organization, see: C. Kummer (1991), Selbstorganisation und Entwicklung. *Theologie und Philosophie, 66,* 547–556.

6   Obviously, this twofold meaning of "substance" explains, why Aristotle thinks the species of things to be eternal. It is only the "second substance", the formal concept of "sphere" or "man", what precedes logically the generating of concrete forms. It is but a core tenet of metaphysical thinking, that these formal concepts represent the principles constituting forms in reality, and therefore must be regarded as ontologically prior to concrete things, as well. In such a way, we conceive the conceptually derived forms of things as "ideas". Whether ideas "preexist" separated from things, as it is common opinion among Platonists, or can only be found as constituents within concrete things, as Aristotle insists on, we need not discuss.

In this way, form can be seen as the real *efficient* cause of the formative process (*Phys.* II 1, 193b 6 ff.), notwithstanding the still valid mechanisms on the elementary level, which Aristotle would classify as "moving causes" (*to kinesan*: Phys II 7, 198a 24). Obviously, the formal cause is not a special factor within or above the elements, but the developing substance as a whole. The entire substance, consisting of and constructed by its elements, determines the interactions of these very elements by its own structural adjustment to a terminal state of perfect realization. Since form in this sense is ontologically prior to concrete structural realization, development must be regarded as an intrinsic teleological process.

As a result, we can state that Aristotle converts atomistic motion into holistic processuality, providing thereby a well-defined ordering frame for the emergence of organized complexity of matter.

3. Aristotelian form concept on the background of modern theories of self-organization

Compared with the thinking of Democritus, Aristotle's form concept must be viewed as a highly important innovation in the understanding of complexity. But today's mechanistic theories yield a far wider range of "moving causes" which explain structural effects in a more sophisticated manner than done by the ancient Greeks. We know of circular causality in modern systems theory; of emergentism and synergistic effects; and, above all, of genetic information as an apparent materialization of the Aristotelian formal cause that seems to be able to direct elementary processes to definite states of form. (Though, more specifically, it might be argued that the very process of sequential coding of DNA silently presupposes a systems theory or some kind of emergentism to generate genetic "information" in an appropriate sense.) Is the explanatory power of these modern concepts of self-organization of matter not strong enough to overcome the metaphysical connotation of the Aristotelian approach?

To answer briefly this question, we state that all these new versions of mechanism just mentioned provide no other concept of causality of form than the one established by Democritus.

Firstly, there is no "circular causality", as suggested by systems theory[7], since a moving or efficient cause is strictly linear, always running from a point A (condition) to a point B (effect) – even in feed-back circuits. It follows from the simple fact that causes always are prior to effects, and never the other way around. So, one can dissect every feed-back circuit in a sequence of A to B events, notwithstanding the ultimate step Z to A which is as linear as the others, showing nothing but the bipolarity of every link in the chain, producing connection as well as being connected. (The only case of circular causality I can imagine is self-activation of gene expression. But even in this case, enhancement of the transcription rate of a gene provided

---

7   Cf. Rupert Riedl (1978): Über die Biologie des Ursachendenkens – ein evolutionistischer, systemtheoretischer Versuch. *Mannheimer Forum 1978/1979*, 9–70.

by its own product is a second step in the course of transcriptional activation initiated by transcription factors coming from other genes. And, moreover, the coding sequences that specify the protein product of a gene, and the regulatory region where the coded gene product acts upon to control its own production rate are clearly distinct regions on the genome.)

Secondly, neither is there any "downward causation", as emergentism suggests – at least not in the sense of a new order of causality. "Downward" refers to an influence from a higher systemic level to the elements at lower levels. So, there should exist special laws valid only at the higher level, which arrange the elements of the lower level in a manner not given by the "moving causes" governing the lower level. Nevertheless, the laws of the higher level must be valid at the lower level too, otherwise a causal influence would be impossible. Consequently, the two classes of laws belong to the same order of explanation, and the so-called emergence offers nothing that could not be analyzed and reconstructed in a reductive way, say, by means of feed-back causation. On the other hand, if we want really new properties to emerge from a certain level of systemic complexity, we have to accept a new kind of explanatory principle transcending the ontological status of the subsystem level, yet comprising it. What kind of laws could constitute such a principle? Is there, for instance, a causality of life exceeding the scope of the laws of nature valid to physics and chemistry? If we think so, we must accept that we are leaving the realm of science, and we are obliged to define exactly the ontological status of this new form of causation[8]. Then I can hardly see how to avoid metaphysics. – Thus, emergentism has no proper explanatory value, but is synonymous with either reductionism or creation.

Finally, it is quite clear today that the genome presents no developmental program in a cybernetic sense, i.e. it is no blue-print of organization, and therefore a theory of genetic preformation is no longer tolerable. The role of the genes is to store prescriptions for the synthesis of peptide chains, to allow a regulated access to that information, and to give the possibility of varying some of those prescriptions at random. But there is a wide range of other mechanisms influencing the effect of developmental instructions, which are not immediately regulated by genetic control. We mention only the interactions of transcriptional (co-)activators and repressors; the RNA-binding proteins affecting nuclear processing of the primary transcripts, and the export of the mature mRNAs as well; post-transcriptional modifications of the genetic message, as RNA-editing and recoding; the active role of the DNA molecule itself as a chromatin organizer provided by histone-dependent twisting and writhing[9]; the inactivation of genes by means of DNA methylation, at least in mammals; the networks of intracellular signal transduction; or, last but not least, the connexin-mediated spreading of second messengers between cells. We should further think of the processes of cell movement which depend not only on the properties of certain adhesive molecules and cytoskeleton elements, but also on the mechanical conditions of the affected cellular layer as a whole. To assert that all these

---

8   This is exactly the problem of the so-called "non-reductive materialism". Cf. A. Stephan & A. Beckermann (1994): Emergenz. *Information Philosophie 22(3), 46–51.*
9   Cf. the model suggested by M. Caserta & E. Dimauro (1996): *BioEssays 18,* 691.

elements and processes result in performing an organism is to neglect the fact that there is always a form the organism already is in possession of and what allows these processes to interact in a constructive manner. To accept this seemingly trivial fact is to agree with Aristotle in the substantiality of the organism, where form governs elementary movement.

The order of all the molecular interactions performing development is, or can be, called "epigenetic coding". The expression ‚epi-genetic' is used to characterize the formative effect of these processes as not programmed in the genes, and the ability of these processes to act upon the selection of genetic information, as well. 'Coding' refers to the structural imprinting of the informational content of a developmental plan. So, "epigenetic coding" in our understanding[10] is the step by step design of a developmental program performed by the processes of protein-gene- and protein-protein-interactions, each step comprising both the programmed state of a hitherto well-organized concert of molecular movement and the programmatic state for further organization. This concept of epigenetic coding, where the idea of genetic information is replaced by the idea that it is the whole of actual organization what informs the molecular processes to produce not only diversification but further differentiation intended for the perfect realization of a definite body plan, corresponds closely with the Aristotelian concept of substantial processuality and inherent teleology.

4. The hydraulic principle and the idea of construction

A main difficulty of Aristotle's substance concept is how to conceive form as *an efficient* cause of the formative process. How can a future state of perfection, not yet realized, act at present as an organizing principle? How can a "real potency" be thought, a concept formulated paradoxically as the "actuality of possibility" in Aristotle's famous definition of motion (*he tou dynamei ontos entelecheia he toiouton*, Phys. III 1, 201a 11[11]). – These are two different formulations of the one fundamental question, how an idea can influence matter. The Frankfurt Theory can help to solve this problem from an empirical starting point.

Of course, the Frankfurt Theory does not deal with metaphysical items. It does not explicitly speak of a formal cause, neither, but introduces the idea of a "hydraulic principle", what, to some extent, touches the same ontological difference between the typology of construction and the reality of organization presented by a concrete living being. To conceive a developing organism as a hydraulic construction implies an identity between the forces and constraints constituting organization and those channeling development into this very outcome.

What we have called, earlier on, the "morphogenetic field" between a perfect state of form and its actual realization, is in this way understandable as the effect of

10 For further discussion of the term, see: C. Kummer (1996): *Philosophie der organischen Entwicklung*. Stuttgart: Kohlhammer.

11 A detailed analysis of this definition is given by: R. Spaemann & R. Löw (^{3}1991): *Die Frage Wozu?* München: Piper, p.57.

a construction type performed by its hydraulic forces. It is because of these forces that we can speak of the objective of construction to be present in a transient constructive stage. The principles of construction provided by these forces are at the same time given with these forces.

Obviously, the hydraulic principle comprises the whole of organization. It is not only the osmotic pressure inside the body together with the tension of its outer cover that furnishes an organism, but the internal connective tissues, muscles and inner or outer skeleton elements as well. All these components are fitted together to perform a structurally and functionally cohesive whole. This coherence of organization precedes and governs the influence of genetic activity at every stage of development, permitting only those molecular interactions and cellular movements tolerated by constructive integrity. Thus, differentiation can be conceived as a process of molecular diversification induced by genetic activity, which does not violate the already established integrity of organization, but by following its mechanical constraints leads to a richer achievement in constructive possibilities. (We must, however, admit a weak point in this biomechanical concept of differentiation: it does not cover the process of molecular pattern formation leading to a basic body plan prior to any change of outer shape.)

That it is possible to understand organismic form solely in terms of hydraulic construction – and this is the claim of the Frankfurt Theory – implies that theoretical explanation of empirical instances of organismic form builds on the same hydraulic principles which govern the molecular processes of morphogenesis resulting in that form. So, as suggested already by the term 'principle' signifying both reason and cause, it is in fact our "idea" of construction what appears at the same time as the actual efficient cause performing it. As this "idea of construction" or "construction type" which makes organismic form intelligible is nothing else but what is traditionally called *causa exemplaris* or *formalis,* the explanation of developing structure given by the Frankfurt Theory of constructive morphology may well serve as an illustration of the Aristotelian concept of causation.

As already noted, the biomechanical concept of the Frankfurt Theory contains an idealistic implication. With a given quantity of living "material", and obeying the same hydraulic principle we can construct quite different organisms, as e.g. a slug, a fish or a newt. So, the type of construction is not given by the hydraulic or biomechanical constraints alone, but answers to the question, what kind of functional type (in particular with regard to locomotion) we have in mind. True, there are different possibilities to construct a certain type of function, and therefore the type of function is insufficient for understanding construction. However, a type of construction is not only a combination of several hydraulic units but something like a technical invention: it represents an idea what an animal should be able to do, or, at least, to look like. These ideas of construction we are accustomed to derive from the concrete appearance of organisms, and therefore call them "Baupläne". So, the Frankfurt Theory *reconstructs* these types of organization by applying the hydraulic principle, and arranges them in an order of deductive morphology, but is unable to explain how these ideas of construction could arise.

Hence, the question asked above, how ideas can influence matter, transforms itself into the even more striking question of how organisms can come to their "ideas". This is the point at which the Frankfurt Theory begins to withdraw and to limit itself to the constraints of a constructivist position by assuming that, above all, it is our intellectual work that constitutes organisms and constructive types, respectively. It is the very intention of the Frankfurt Theory to re-establish morphology as a science by basing it on principles more suitable for this objective than Darwinian adaptation. But the objects of constructive morphology are treated by the Frankfurt theory as constructions *per se* which are derived and elaborated by expressions drawn from every day's experience (like, e.g., 'machine', 'hydraulic pressure', 'tension', etc.). Thus, the validity of these constructions is only proven by explanatory usefulness: regarding organisms as hydraulic machines helps to handle organisms as objects of a certain scientific interest, but it disregards the question whether organisms are so constructed *in reality*. It seems reasonable to deal with the idealistic impact of constructive morphology in a manner that avoids metaphysical implications, regarding the ideas of construction only as regulatory principles that systematize our understanding of living forms. However, in our opinion, this strategy of reducing teleology inherent in organismic construction to constructivism will not work.

First of all, as a matter of fact, biologists yet remain interested in the reality of their objects, that is, of living "systems", and this implies the indispensability of our understanding of the real morphogenetic processes. Though this may seem a hopelessly naive epistemology, it is the most common one among biologists and the one which motivates their work. So, to ascertain that organismic form is something particular the mechanistic issue of biology is unable to explain, and yet to grant molecular biology its own right to explain the development of form is to introduce a duality into the methodological unity of the biological subject. It is therefore important to regard organisms as subjects to themselves, that is, as their own agents, and not as products constituted by an alien, systematizing, subject. This characteristic feature of biological questioning must be maintained within the epistemological framework of constructivism, as well.

Secondly, constructive morphology itself understands its hydraulic principle as a real, biomechanical, factor of development and not as a mere theoretical model of our morphological deductions. As we have seen, the effect of the hydraulic principle is that of a formative coherence of the whole organism, bringing the elementary "movements" into a concerted direction and so acting as a formal cause. Yet, the constraints of the hydraulic construction as well as its shape-giving conditions and the tethering of its elements work on the level of morphogenetic forces. So, they positively influence the molecular order and must be conceived as true mechanisms (which is the meaning of 'cause' in natural sciences) of the developmental process. Nevertheless, the question remains, whether the hydraulic principle, doubtless being a *conditio sine qua non* of development, is already the *sufficient* reason explaining the different types of organismic forms. Anyway, to introduce the hydraulic principle as an answer to the question, why an animal shows a certain shape, is to put it on the ground of concrete developmental mechanisms, and not to leave it in the heights of pure theoretical consideration.

A last point concerns the constructivist issue itself. According to this issue constructive morphology claims to "construct" its objects as hydraulic machines. However, in every day's speech we would not say that we "construct" organisms as machines, but that we *compare* them with machines. And this is not exactly the same. To compare two things means that there is a *tertium comparationis*, realized in both, which is the reason for using one item, already known to us, to explain the other. So, comparing organisms with machines implies a certain common feature within both that helps to understand the unknown object, the organism, by means of the already analyzed one, the machine. Obviously, this common feature is the purposefulness of structure. However, to approve this way of understanding living nature by comparing it with technical products is to accept the reality of structurally expressed ideas in both, no matter where those ideas came from. That means to accept teleology in nature in a strict sense[12]. Then, it is no longer tolerable to put the idea of construction only into the mental process of our reconstructions instead of localizing it in the complexity of the organization itself, like the constructive idea of an engine does not exist in the brain of its inventor alone but can been seen in the construction, too. And that is exactly what Aristotle suggests.

Maybe, we have no other way to explain the structural teleology of organisms but by means of mechanistic interpretation. Then we have to accept a world where it is quite normal that the informational content of a whole book emerges by chance. Or, we deny the teleology of nature completely, as Darwinism does. But if we think that constructive morphology describes organisms in the right way, we have to acknowledge its ontological impact.

---

12   The ontological background of the structural parallelism between technical and natural constructions is clearly worked out by: H.-D. Mutschler (1996): Über die Möglichkeit einer Metaphysik der Natur. *Philosophisches Jahrbuch, 103,* 9–13.

The butterfly and the lion

Giuseppe Sermonti

1. Rise and decline of the "genetic" paradigm

"Evolution" is widely considered as the key and the lustre of modern biology, which is commonly dated back at the formulation by Charles Darwin of the selective theory (1859). As a matter of fact, evolution is usually alluded to, but rarely the object of actual scientific investigation.

Population genetics, which was considered for a long time "evolution" *tout court*, turned out to be rather a branch of ecology, with negligible impact on speciation and phylogenesis. Even comparative molecular anatomy did not provide any important clue to the mechanism of evolution and to taxonomy.

The real beginning of modern biology and the foundation of the present view of life are almost a century old and correspond to the rediscovery of the Mendel's rules and to the acceptance of the Weismann's germ-line theory. These events produced the eclipsis of the dynamic evolutionary paradigm and the onset of the static "genetic" paradigm. A central self-reproducing agent was assumed (cell nucleus, genotype), generating the organism as a side-product, and defended from the environment by the so called "Weismann's barrier". The "Central dogma" of molecular biology was a restatement of Weismann's theory, with DNA in place of germ-line and proteins in place of phenotype. The Weismann's barrier, based on weak experimental evidence, became perentory on chemical ground. Random mutation of DNA took the role of the exclusive transformational agent, but mutations backing specific morphological differences as still wanted, and we are still in quest of the very mutations which separate the butterfly and the lion.

After the discovery of the DNA's role, the amount of DNA per nucleus was made responsible of progressive evolution. Protists contain in fact nuclei made up of nucleotide pairs of the order of millions and eukaryotes of the order of billions. As it was soon realized, this difference did not concern the amount of genetic information as much as the structure of the nucleus, very parsimonious in protists, with genes in a continuous series, while redundant in eukaryotes, the genes of which were largely spaced and internally split. Doubts were expressed about which structure was more primordial and ancient. Calculations made possible both in protists and various eukaryotes made clear that the total number of genes (or gene types) per haploid nucleus was substantially constant (of the order of 5,000) in bacteria, flies and amphibians. Even in mammals, the provisionally estimated number of genes was of the same order of magnitude as in bacteria, surely not of such a magnitude as to justify the enormous disproportion in size and complexity. Almost 90% of mammals' enzymes are detectable in prokaryotes.

Detailed investigation of the genetic code and sequences of DNA in the most disparate living beings produced really shocking results. They provided the definitive evidence of the universality of DNA code and structure, but were not such as to account for the paramount morphological differences, among types, orders or families, such as that from the butterfly to the lion. The "genetic paradigm", aiming at the explanation of the external differences on the basis of internal gene libraries, failed. Its failure was a second defeat of the "evolutionary paradigm", which had trusted to genetics the interpretation of the organic transformations. As a matter of fact the genetic make up of the species was disappointingly persistent, in spite of the impressive variety of the biosfere.

2. Essential constancy of DNA

When the genetic code was deciphered, in the fifthies, it was not a surprise to anybody that it was the same in all living beings, from virus to elephas. Biologists had already guessed that the basis of life should have been compulsory and unitary. This was considered as evidence of an unavoidable chemical determinism at the basis of life. It was A. Sibatani, in Japan, that remarked the "arbitrariness" of the genetic code. There was not stereochemical law that could have predicted the correspondence between given nucleotide triplets and specific aminoacids. The rules were established by "natural conventions", with no obligation, but, once established, they could no longer change, as a "conventional" language cannot change once agreed upon, at the risk of a general misunderstanding. In more physical terms F. Crick defined the stabilized genetic code as a "frozen accident".

The genetic code had proven the *unity* of life, the link among all living beings. It was now the reading of the messages written in the universal language which should have accounted for the *multiplicity* of the biosphere. This was explicitly forecasted by some geneticists, who declared that molecular changes in proteins could give rise to a gradual differentiation among individuals and eventually produce new species (I.N.M. Lerner, 1968). Mutations, underlined G. Montalenti (1965), are the most important (if not the only) source of variability to the effect of specific differentiation.

After the deciphering of the genetic Hieroglyphics, the comparison among genetically unrelated species became a possibility. On the basis of the comparison of citochrome c primary structure from a number of organisms, the conclusion was soon reached that the more two species were taxonomically distant, the more were their homologous macromolecules different. But were these differences "responsible" of the taxonomic divergencies? Or just markers of evolution along independent lines (as decaying radioisotopes in rocks)? The latter turned soon to be the case. The differential aminoacid residues in homologous molecules were substantially *neutral* and unsuitable to foster phenotypic differences. As F. Jacob wrote (1977), "Biochemical changes do not seem to be a main driving force in the diversification of living organisms. (…) It is not biochemical novelty that generated diversification of organisms. (…) What distinguishes a butterfly from a lion, a hen from a fly, or a worm from a whale is much less a difference in chemical constituents than in the organization and the distribution of these constituents".

Geneticists were convinced that, by exploring the core of the cell, the *sancta sanctorum* of the molecular mistery, one should have approached the key of differentiation. This was not the case. "The more one approaches the molecular level in the study of living beings, the more similar thay appear and less important the differences between, for instance, a clam and a horse become" (Dickerson, 1972).

This was a disappointing conclusion. The gene differences among taxa resulted accidental, irrilevant and insensitive to natural selection. The difference in (only) the 66th residue of the cytochrome c chain between man and chimp (an isoleucin instead a threonine) was just a casual mark and had nothing to do with the variance between the two subjects (the same residue was equal – threonine – in the whale and in the worm).

One could conclude that 'molecular evolution' did never occur to any important effect. As Sir Peter Medawar (1967) wrote, "In a certain important sense all chemical evolution in living organisms stopped millions of years before our faintest and most distant record of life began. So far as I know, no new *kind* of chemical compound has come into being over a period of evolution that began long before animals became differentiated from plants".

3. The zootype

In the absence of convincing evidence of evolution and selection within structural genes, genetic evolutionary research started exploring such differences among "regulatory" genes. These were the new candidates for explanation of divergent evolution. Recent work in developmental biology has shown that there is indeed a class of gene that specifies relative positions within the body of several unrelated animals. Mutations in such genes produce ample shifts in morphological characters. The *Hox cluster genes* are the best known of these. The first Hox cluster type (Homeobox containing) genes were discovered in Drosophila. When their occurrence was found in other animals (hydras, nematodes, amphioxus, leech, frogs and mouse), it was soon proposed that the Hox cluster would become a Rosetta stone in the study of animal development and evolution. Expression of the genes at the 3' end of each cluster commences at anterior body levels of the embryo, and for each gene moving along the chromosome in a 5' direction, expression commences at a more posterior level.

The secret place of taxa differentiation appeared within reach. Hox cluster genes turned out however to code relative position within the organism rather than any specific structure. The Hox cluster genes were virtually identical in all tested animals! For their general occurrence in all animals they have been called "zootype" (Slack et al., 1993). They define animals, but they are not different between different types or classes of animals, 'between a worm and a whale', 'between a clam and a horse'. It was assumed that this system is very ancient and was in place in the common ancestor of all modern animals. The zootype has persisted ever since, perhaps because mutational change to it would be too distruptive to the developmental program of a plausible animal. It was as universal and constant in animals as the genetic code!

4. Modulation of DNA

The situation was apparently the following: a virtually constant DNA information was shared by the most disparate organisms. Such information was not however equally exploited in different cells or tissues of the developing organism, non in different species of a developing phylogeny. In other words, substantially the same DNA is available in all organisms and in their different parts, in spite of the occurrence of large morpho-physiological differences among organisms and among their organs and tissues. DNA is a kind of silent violin producing different tones according to which region of its strings is excited, i.e. to which genes occur to be transcribed.

This situation appears particularly striking when one compares successive stages in metamorphic animals, different castes in social insects, alternating gametophyte and sporophyte in plants. The egg, the 'worm', the crisalids and the adult of a butterfly are impressively different from each other, yet they share an identical genotype. When the larva becomes chrisalid not a real metamorphosis takes place, rather most of the larval organs become destroyed and the adult organs develop from particular groups of cells, already present in the embryo, known as imaginal discs. The process is regulated by the molt hormone ecdisone. The new being is deeply different from the disappearing animal from which it emerges. In social insects as termites (Noirot, 1982) large differences occur among the castes of soldiers, workers and winged sexuates (king and queen) irrespective of the genotype. In higher plants gametophytes (pollen and ovary) and sporophyte are uncomparable, yet their differences cannot be accounted to by their genomes or their ploidy.

One could ask: why, to explain differentiation, we insist looking after DNA mutations, when gigantic differences are under our eyes, established in the absence of any gene change. The elisir of variability is surely under the heading of epigenetic manifestation of one and the same genome, the role of gene mutation being possibly only that of phenotypic stabilization.

Some topologically distributed force (a field), foreign to DNA itself, is involved in the 'evocation' of different gene activities. DNA looks thus not as the agent of differentiation, rather as a kind of constant substrate on which some formative causation exerts its operations.

Differential gene transcription is for sure involved in metabolic differentiation. This was shown to be connected in bacteria with different isomeric states of DNA, dictated to DNA itself by forces apparently foreign to the DNA message. DNA transcription by RNA polymerase is dependent on local DNA denaturation and uncoiling. Prokaryotic DNA is largely involved in elastic supercoiling known to play a prominent role in genetic transcription. Such supercoiling, or torsional stress, is mediated by enzymes called topoisomerases, namely producers of different topological isomers of DNA (Wang, 1985).

Until the eighties topoisomerase-induced torsional stress was thought not to occur in eukaryotic DNA (Gellert, 1981). Elastic supercoiled DNA is more accessible to nuclease and preferentially sensitive to digestion with Dnase I. By that digestion it was possible to show that some torsional stress is present in eukaryotic DNA also (Villeponteau et al., 1984). The restricted regions sensitive to digestion with DNase I are those comprising the active genes. An active gene is ten times more accessible to digestion than quiescent DNA. The regulatory region of the gene is hypersensitive, being digestible up to 100 times as much the inactive DNA. Effective transcription of genes is only possible in the presence of superhelical tension. When DNA supercoiling is released and DNA is returned to a DNase insensitive ground state, gene translational activity ceases.

An important problem at this point is to what extent chromosome topology is stable and inheritable. It is known that torsional stress is retained in specific genes present in differentiated tissues after they have ceased their translational activity and that self-reproduction of torsional stress occurs in some virus. Is thus not a different DNA message which is responsible of cellular and tissue differentiation, rather it is the isomeric variation of a constant message. How this situation holds in phylogenesis is an open problem.

We have represented DNA as strings of equal violins differentially excited in time and by various musicians. The question remains: what does direct the violinist? Recourse to some sort of 'morphogenetic field' is compulsory at this point, although its nature remains elusive so far.

5. Self-organization

To the effect of differentiation the Central Dogma has been, to some extent, reversed. It is no longer a self-reproducing DNA which dictates the differential transcription of proteins; on the contrary, some special proteins, (polymerases, repressors, hormones, maturasis, topoisomerases etc.) dictate to DNA when and where to work. Small molecules control in turn the activities of these proteins. The end result of these processes is the phenotype. As previously shown, quite different phenotypes may emerge from one and a single DNA. No Weismann's barrier protects the state of activity of DNA, which is, on the contrary, demanding some external prompter to keep or modify its differentiated state. Physio-morphogenetic differentiation is entrusted not to the primary structure of DNA, but to proteic activators acting on DNA. The time and place of their action is the very problem of developmental biology and evolution. DNA, its activator proteins, as well as the small molecules affecting them, all are poorly specific. We are confronted with a problem concerning topology and chronology of some directing forces (assumedly electromagnetic) which are embedded in the organism. As an extreme case of a small molecule producing taxa differentiation, one can recall the case of axolotl and salamander (*Ambystoma mexicanum*), formerly classified in different suborders, later shown as belonging to one and a single species, at different metamorphic stages, exhibiting different phenotypes in virtue of a variable concentration of iodine in the water.

The only safe differentiators among various organisms are eventually the organisms themselves. In some way they transmit their shape to the progeny, be it through the cytoplasmic structure of the egg or its cortex, or as maternal effect. This condition is well expressed by the term "self-organization".

6. The problem of stability

We have presented the primary structure of DNA as substancially constant to the effect of both differentiation and evolution. As a matter of fact the genome is affected by quite a few alterations, which some theorists have elected to evolutionary factors. We have been hearing a lot about such things as mutation-rates, hot spots, jumping genes, transversal inheritance, genes in pieces, redundant genome and recombinant DNA; and we should expect the species to be in a permanently unstable state. And yet, the world of biology seems reasonably stable, and repetition is the norm in ontogenesis and stasis in phylogenesis.

Biology, growing up in the thrall to macromolecular genetics, has glibly postulated an immediate relationship between genotype and phenotype, where the temptation is to look on the latter as largely superfluous and a little more than a serviceable tool in the quest for the genotype.

Formulations as "one gene – one enzyme", "one triplet of nucleotides – one aminoacid", not to mention Dawkins' "selfish gene" have reduced the organism to a vague walker-on against life's molecular backdrop. Since the "directive central agent" is substancially restless, one should forecast an equivalent phenotypic variability, as actually favourite by selectionists aiming at a primary source for evolution.

Against such expectations, the experience of genealogists, morphologists and embriologists attests a fundamentally stable organism, the phenotype variability being largely dissociated from that of the genotype. This is not only due to the fact thet most genetic variations are neutral, recessive or hypostatic, rather to their composition and adjustment within the complex pathways (epigenesis) leading from the primary genetic instruction to its ultimate outcome in form and function.

Organisms have a peculiar property which we in Italy call "Gattopardesca" (leopardian), from the novel "Il Gattopardo" by Tomasi di Lampedusa. The book, as sometimes happens, is largely known for a sentence: "To ensure that things remain as they are, we must expouse the change taking place anyways". This is no piece of hypocrisy or chamaleontism: it is a theory of stability. And this is how Nature works, for immutability and rigidity spell to death of a system: to go on living demands the art of adjusting.

A bicycle remains upright and goes forward in a straight line because is continuosly wobbling and straightening up again, imperceptibily leaning now to one side, now to the other. We beside the tightrope walker who stops wriggling!

Authors of this century – Schmallhausen, Wright, Lerner – advanced this paradox of biology; namely that genetic variability has a "stabilizing" function on phenotypes. A population presenting numerous genetic polymorphisms with the same end result for the phenotype is better able to resist pressure to introduce changes in the developmental process, in comparison with a monomorphic population. By analogy we could say that if the organism has a biochemical redundancy for performing a certain function it will be the better able to preserve it in the face of genetic or environmental pressures. The organism has an advantage where stability is concerned if it can "canalize" a plurality of streams toward a limited number of outlets (Waddington). Under conditions such as these, pressures from the environment operates to stabilizing effect. Species taken out of their wild environment and placed in captivity manifest an unexpected explosion of morphological instability. Belajev (1981) illustrated this situation by reminding us what happens with his ranch-bred silver fox.

This tendency to channel different processes in specific directions was suggested by Hans Driesch (1914) in his "equifinality" principle – i.e. a particular point of morphological arrival can be reached in the developing embryo, or in the organism undergoing regeneration, from different points of departure and following different development paths.

The propensity of classical genetics to "flatten" the phenotype on the genotype, to the point where the two merge (the outcome of decades of research on the most elementary organisms – moulds, bacteria, viruses) has had its day. We now know that, in order to have a quiet life, an organism must be careful not to resemble its genotype too close and not to induce it to reveal all too readily its vagaries and its weakness.

A non-genic continuity of the organism is postulated, largely referred to a self-reproducing morpho-field, to which the biological stability is entrusted. A question can be rised at this point: are there in the cell structures which reproduce or modify themselves irrespective of the working of DNA?

7. Coherence domains in the organism

Several types of polymeric structures occur in the cell, able to self-reproduce only when starting from similar structures. Central among them are the membranes, which only develop as extension of preceding membranes , from proteic materials instructed by DNA. A biologic membrane is in a state known as *liquid crystal*. This consists of rod-shaped molecules having a positional as well as a specific orientational order. Their alignment, which extends over millions of molecules, is made stable by an attractor called the "director" and may have a "chirality". Both alignment and chirality are transferred to neighboring molecules through an effect at a distance, extending a "coherent domain" and comparable to the "morphic resonance" of Sheldrake. The liquid crystals provide a good example of the dynamic "coherent regime" of the matter field, where particles "lose their individual identity, cannot be separated, move together as performing a choral ballet, and are kept in phase by an electromagnetic field which arises from the same ballet" (E. Del Giudice, 1993).

A dramatic example of a coherent regime changing its state not by effect of a DNA mutation, but as a result of a proteic intruder is the well known case of the "mad cow". A pathologic protein, called "prion", affects a coherent regime in neuronal cells eventually producing a lethal disease, the "Bovine Spongiformis Encephalitis", or BSE. In fact we are not faced by a proteic inheritance. A normal protein in a coherent regime (PrP-C) is arranged in alpha-helices, while the pathogenic one (tre 'prion', or PrP-Sc) is partially folded in a beta-sheet structure.

According to Stanley Prusiner the diffusion of the anomaly depends on an action by contact. When the normal protein is unfolded in the cell, the occurrence of an anomalous prion would cause it to switch into the beta-sheet structure. A chain-reaction would diffuse the anormal structure to the normal proteins, eventually producing the illness. This process, which has been called the "rotten apple effect" was proven beyond any doubt. A radioactive normal protein (PrP-C*) was mixed in vitro with an unlabeled prion (PrP-Sc). Two days later a labeled anormal protein was detectable (PrP-Sc*). The initially normal protein adopted the 'rogue' form by contact with the bad one.

In the absence of the intruder the PrP-Sc protein is kept in a coherent regime and self-organizes its normal lay-out. When it intervenes a new coherent regime is established, again self-reproducing, lethal in case of PrP-Sc prions, perhaps only different in other instances.

Liquid crystals (to which I assume prions to belong) are well known to be sensitive to electromagnetic fields, adjusting the orientation of their director and consequently their pattern. They are reasonable candidates to the role of molecular imprintings of forces arranged in morphogenetic fields. Perhaps their organizative modalities may concur in distinguishing, to use the terms adopted by F. Jacob, "a butterfly from a lion, a hen from a fly, a worm from a whale".

Literature

Belajev, D.K. (1981):Destabilizing selection as a factor of domestication. In: "Genetics and the well being of humanity". (in Russian) Nauka, Moscow: 53 – 66.

Del Giudice, E. (1993): Coherence in condensed and living matter. Frontier Perspectives, 3/2: 16 – 20.

Dickerson, R.E. (1972): The structure and function of an ancient protein. Scientific American, **226** (4): 58 – 72.

Driesch, H. (1914): The History and Theory of Vitalism. MacMillan, London.

Ho, M.-W. (1993): The Rainbow and the Worm, the Physics of Organisms. World Scientific, Singapore.

Jacob, F. (1977): Evolution and thinkering. Science, **196**: 1161 – 1166.

Lerner, I.M. (1968): Heredity, Evolution and Society. H.W. Freeman & Co., San Francisco and London.

Montalenti, G. (1965): L'Evoluzione. Einaudi, Torino.

Noirot, Ch. (1982): La caste des ouvriers, element major de succès evolutif des Termites. Rivista di Biologia, **75**: 157 – 196.

Slack, J.M.V., Holland, P.W.K., Graham, C.F. (1993): The zootype and the phylotypic stage. Nature, **361**: 490 – 492.

Villeponteau, B., Lundell, M., Martison, H. (1984): Torsional stress promotes the DNAse I sensitivity of active genes. Cell, **39**: 469 – 479.

Wang, J.C. (1985): DNA topoisomerases. Annual Review of Biochemistry, **54**: 665 – 697.

Organism – Ecosystem – Biosphere:
Some comments on the organismic concept

Harald Riedl

Introduction

Apart from living organisms, the term "organism" has been applied to various systems at a higher level by various authors. A famous example of recent times is Lovelock's Gaia concept, according to which Earth as a whole including its flora and fauna as well as its atmosphere and its interior is regarded as an organism, in which everything is interacting with everything else. Closely related to the concept of organism is holism with its basic assumption that the whole is more than the sum of its parts and cannot be properly understood by purely analytical procedures, therefore.

In the mid-thirties a dispute arose among ecologists about the nature of so-called "biotic communities", which were regarded as "complex organisms" by Clements, Philipps and their school. As a result of this dispute, new concepts were developed such as that of the ecosystem by Tansley (1935), which are widely used until today.

The sum of terrestrial and aquatic ecosystems is usually called "biosphere" using a term introduced for the first time by E. Suess in 1875. Contrary to Gaia it is confined to Earth's surface only. Though it never has been looked upon as an organism, it may be used in relation to the other terms discussed here as one of the few concepts that has never been misinterpreted or questioned until now at least to my knowledge.

New aspects introduced by physics and mathematics may exert a strong influence on future developments in this field offering unexpected ways for defining terms used for a long time for their surface value alone. "Feed-back" and the related concept of homoeostasis, or the whole recent development of open systems thermodynamics have proved useful already to understand processes within individual organisms as on a social level. One of the modern aspects introduced mainly in molecular biology until now and not sufficiently understood in its full meaning by many biologists includes information and communication as physical properties applicable to many biological problems in quite different areas. It will be used here to help formulate a new theory of organisms on different levels and their interactions beyond the undeniable success of general systems theory.

Living organisms

Living organisms as the prototype for any kind of organism are defined by a number of properties, all of which apart from the phenomenon of life as we know it are necessary attributes for any system to be called organism. The applicability of the term "organism" to any system has to be tested against a living organism, therefore. I do not want to give a formal discussion of all previous views on the subject, as this would lead too far, but confine myself to a few statements which I adopted from many different sources. As far as I can see, Kant has been the first to deal in detail with the concept of organism, and has said nearly everything necessary in his "Kritik der Urtheilskraft", though his language may seem strange and too complicated to a modern scientist.

    The common properties of all kinds of living organisms may be listed in the following way:

1) A living organism is a whole on its own particular hierarchical level in the realm of nature and the universe. As it is open in respect to higher levels which include it as a part, wholeness can be ascribed to it only in a restricted sense. This situation is reflected in a special term applied to it, that was coined by Koestler (1967, 1978), who called such relative or incomplete wholes "holon".

2) A living organism can be compared to or spoken of as an elementary particle in the sense that it is the smallest material unit displaying all the qualities of life. A detailed discussion of this topic newly introduced by me here will follow in a separate paper ready for the printer.

3) Organisms are self-regulating systems communicating with their environment, from which they take the energy necessary to keep up a steady state (homoeostasis) for a given time.

4) The concept of organism is inseparable from organisation as a prerequisite of its functioning as a whole. Organisms are structured, that means composed of smaller interdependent elements distributed in space and time according to a special design by which they are discerned from organisms of a different kind. All these elements act in a coordinated, "organized" way, that was called "innere Zweckmäßigkeit" by Kant. This is possible only if we assume the existence of a central control mechanism kept going by feed-back reactions. It is one of the great mistakes in the otherwise convincing theory of self regulation that it does not sufficiently take account of the structure that is to be maintained as a whole together with its development in time, and the whole as the central control mechanism. It is true that morphology as the expression of the whole is complementary to the various physiological processes involved in self-regulation. Once I compared adult "Gestalt" as complementary to the sum of functions to a standing wave (Riedl in prep.). It is coordinated function that calls for the whole as central control mechanism. Throughout individual ontogenetic development one process triggers the next, but even these triggers have to follow an innate design that may be incumbent in the genome as a whole corresponding to the wholeness of the adult organism, but certainly not in particular genes. That notions like the preceding one are not generally accepted seems to be the result of dogmatic denial of direction towards an unknown aim to evolution in general in

the course of which the various organisms are designed as necessary steps by and within a greater whole, and to the fear of any connection with teleology as an "unscientific" approach.

5) Organisms have distinct boundaries in space and time, so that exterior and interior forces can be discerned which regulate the processes that lead to alternating periods of change (development) and homoestasis. These boundaries delimit what is generally called their peculiar "Gestalt" and may be extended into time as "Zeitgestalt" in living organisms.

It may be noted here, that there may be problems of drawing the boundaries of what is an organism as soon as we relate the term to that of the individual. There are well known examples like that of corals and other primitive Metazoa, in which the question may arise, whether the individual coral or the colony as a whole correspond to our concept of organism. Very close symbiotic relationships such as lichens also may be considered as an organism, though they are certainly not individuals. We shall return to that question later on.

If we assume that an organism as a whole cannot be understood by studying its individual parts and their properties, because it acts on a higher level displaying qualities not found on the lower, the question arises whether we do not ask for something impossible. Many authors such as Cramer have expressed the view that there is no scientific method to study wholes, as any scientific procedure starts with analysis. This is certainly not true for living organisms. As we have seen, organisms are wholes only on their own level. Seen from a higher level, they are parts of a greater whole and can be understood by their interaction with other parts using analytical techniques adapted to the particular problem.

Organisms are considered as parts of an ecosystem, in which they are performing certain functions. This may be one of the reasons, why there is a strong tendency of scientists sympathizing with the holistic approach to regard ecosystems as organisms of a higher order. Let us examine this proposition in detail and also try to find a satisfactory definition for the term "ecosystem".

Tansley (1935) introduced the concept of "ecosystem" in an article in which he tried to give a theoretical basis and terminology to the young science of ecology in answer to a similar attempt of Phillips published in the same year. Phillips talked about "biotic communities" as "complex organisms" following F.C. Clements, who had expanded the theory of successions in vegetation and introduced the notion of "climax" as the final aim to which succession leads in many publications from 1901 onwards. Contrary to many other scientists both Clements and Phillips had fully integrated time in their argument about the true nature of biotic communities and thus had correctly interpreted organisms as four dimensional entities. Phillips insisted, that succession in the strict sence is inherent in the biotic community consisting exclusively of the living organisms present at a particular habitat and is not determined by abiotic factors in any way, though their importance is not denied. Succession is interpreted as development of a complex organism terminated by climax as its adult stage. The holistic point of view in the sense of Smuts is explicitly adopted.

Tansley, on the other hand, did not accept holism in scientific argument. He criticised the term "biotic community", as plants and animals are too different from each other to be able to form anything like a true community. From our present knowledge, this is hard to understand. He is certainly right, however, when he expresses the opinion, that biotic and abiotic factors together have to be considered in any study of social structures including different species and their development in nature. For the complex of habitat and living organisms he coins the term "ecosystem" in order to stress their interaction. This interaction is not limited in time and has no direction inherent in the so-called community. From this argument it is obvious, that an ecosystem cannot be regarded as an organism of any kind.

The term "ecosystem" has met general acceptance and is widely used now not only by the scientific community, but also by environmentalists and politicians all over the world. What is its true meaning if we accept Tansley's opinion, that it is not an organism of any kind (and, I am afraid, we have no choice but to follow his argument at least in several points)? His definition is as widely accepted as the term itself. Odum (1973), for instance, in his already classical treatise on ecology repeats Tansley's concept and illustrates its application. Nevertheless, in my opinion it is not satisfactory, as the parameters used in the definition are not sufficient to describe any particular phenomenon in nature. Let me give an example: you are walking through a deciduous forest here in central Europe. Along your way, there are several small outcrops of limestone, on which moss and lichens and several small flowering plants are growing. There are stumps of old trees also covered by mosses, lichens and a few fruiting bodies of fungi. In the decaying wood, insects, myriopods, snails and other small animals are active and fungal mycelium can be found everywhere. A path is winding through the forest, and along its margins a quite different community of flowering plants, some of them typically ruderal, has established. In the deep furrows of the path water has aggreggated and an ephemeral community of microorganisms and a few small animals has established. What then is the ecosystem: the forest as a whole, or the several different habitats within it? It is generally assumed now, that they are subsystems within a greater system. On the other hand, we often hear talk about the "forest ecosystem". In this case, a general category is meant and not any particular forest. This general term does not inform us about any subsystems and even less of the various components. Is ecosystem a particular habitat with all the organisms living there we can find on a map, or is it a general category? The term has been used in both ways, and from Tansley's definition one cannot say that one application is better founded than the other.

This is a typical example of the weak terminological basis, on which many modern biological concepts are founded, especially but certainly not only in ecology. Useless to say, that the definition does not allow the application of the organismic concept on ecosystems, as they are not clearly delimited in space and time, do not act as a whole, are not self regulating under outward pressure but just changing their components by degrees, and do not show a central control system or a strict internal organisation as a whole. I am tempted to say that they do not correspond to factual reality.

If we want to maintain the term "ecosystem" in spite of its obvious emptiness and inflationary use, we can give only a very wide definition to be used pragmatically such as the following: We may call ecosystem any local complex involving living organisms and abiotic elements we want to study in a particular context if it seems desirable. Whether such a generalization still can claim any heuristic value has to be tested by practical application.

What is real within so-called ecosystems are the relationships among their various components, especially the relationships among a limited number of organisms. They may follow very strict rules in the sense, that certain species will associate only with certain other species. Examples are mutualistic symbioses, parasitism, prey-predator relationships, certain kinds of saprophytism, etc. The best way to classify this kind of relationships to my knowledge is the revised version of Starr's system published in 1975. There is a series of increasing interdependence that finally leads to a more or less organismic organization such as in lichens. The lowest degree of interdependence is that known as commensalism. We usually underrate, however, the deterministic effects of decay of certain organisms and the involved microbes or small animals on changes within a community, in which commensalism seems to prevail. To study relationships of organisms is a way to describe them as wholes on their own level but acting as parts within a higher order.

Species

It seems necessary here to mention, that for living organisms there is still another greater unit they form part of: the species. Species inhabit a certain area in consequence of their history and their ecological requirements, but are primarily connected with the time factor as members in a genealogical sequence. Moreover, among members of a species there is a system of information exchange that does not work between species. Communication acts on one level only among individuals with equal rights, but there is no central control mechanism creating a hierarchical system of information exchange. Though they have a distinct extension in time from their first appearance to their extinction, such fixed boundaries are lacking in space. They certainly are not self-regulating. They are units, not wholes.

Biosphere

All the terrestrial and aquatic ecosystems in the widest sense comprizing all the species and habitats existing at a given time taken together form a mosaic covering almost the entire surface of our planet. This outermost layer on Earth's surface is called biosphere, a term originally coined by the Austrian geologist Eduard Suess in 1875 and later on taken up and made popular by the Russian geologist Wernadskij. The biosphere is the stage, on which evolution of life on Earth has taken place. It occupies a distinct segment in space and time. Whether we take it as something separate from the deeper layers as usually is done in the life sciences or as part of a unity including the whole planet as is done in the Gaia hypothesis of Lovelock is of minor importance in the present context. I am inclined to prefer the latter version if some amendments are made, but for our argument biosphere certainly is the most important region.

The concept of climax is heavily criticized in modern ecology, as it is changing all the time with changes in the climate, and by necessity must change over long periods, whenever its components are changing in the course of biological evolution. Succession includes comparatively short term changes within the changes going on at a much larger scale. If succession is development at particular places during a particular time, it is only like a wave in the ocean of evolution, that means permanent change as long as our planet exists and is inhabited by life.

Is biosphere, or rather Gaia a living organism of a higher order? If it is, its life as well should be different from what we normally call by that name, so that Gaia again would be an elementary particle as smallest unit representing planetary "superlife". Is it a whole? Lovelock has given many examples of its self regulating capacities. It has its boundaries in space and time as a four-dimensional entity. That it is well organized we see as soon as the order is disturbed as just now in the environmental crisis we are passing through. There is only one more question we cannot easily answer: is there a central control mechanism? If it is an organism, evolution has to follow a direction and cannot be interpreted by pure chance. This is contrary to the nowadays prevailing neodarwinistic modes of thinking, but necessary, if we attach the notion of "Zeitgestalt" to the definition of organism. Lovelock avoided to consider these consequences and never became quite clear, whether he regarded the Gaia system as similar to an organism, that could be taken as a model for it in many respects, or whether he actually took it as an organism of a higher order. As he still clung to neodarwinistic concepts, only the first interpretation is possible. Present science cannot solve this question and the answer is still a matter of personal faith. Nothing is proven – not even neodarwinism in spite of the assertions of its adherents to the contrary.

Information

There is still another way to approach our problem that may turn out to be less controversial and yet be useful in our search for an answer. This approach makes use of modern information theory.

Information in the sense of physics is potential and closely connected with mathematical probability. It is the opposite of determinism. The German physicist F.-A. Popp (1986) differentiated between potential and actual information. Actual information is what we can perceive with our senses, it is quantitatively of a much higher order than potential information, that cannot be perceived but is present as a sum of possibilities that may be recognized only under certain conditions. Genetic information goes far beyond what is expressed in the phenotype, for instance. It may be considered as dormant for its greatest part.

The greater the information content in any system, the smaller is the deterministic component, the more probability takes its place. A highly specialized species of animals or plants has a comparatively low information content. High information content results in or is equivalent to greater plasticity.

Living organisms in general display a high content of potential information compared to abiotic systems. This is best expressed in animal behaviour. Plants have a less fixed morphology, their potential is expressed in individual morphological differences that are strongly influenced by environmental factors. In all these reactions, potential information is canalized by forces from within the organism and from the environment in the process of getting a visible expression. If the content of potential information were low, they could not respond to changes in the environment. Self regulation is likewise the result of information present. In this case, homoeostasis is maintained, there is no actual expression recognized as change.

Classical physics is strictly deterministic. For one particular cause there is one particular effect. In quantum mechanics, the concept of probability has been introduced by Heisenberg's famous uncertainty principle.

In an ecosystem we have abiotic and biotic components. Reaction to environmental change is determined by the information present in these various components that react individually as a consequence and not as a whole. This is different in the purely biotic relationships such as symbiosis, etc. For any such relationship it is necessary, that the partners have at least some informations in common or that informations correspond to each other. The sum of these informations can be considered as the information content of the system as a whole and may produce at least some organismic qualities in the system.

Potential information contained in the biosphere is the sum of all information that is and has been present in all the organisms on Earth throughout the whole course of evolution. Whether there is any additional information connected in one way or another with the physical environment we don't know. It has always been a matter of wild speculation and goes beyond the limits of present day science. It is not known, moreover, whether this total body of information is more than the sum mentioned above. We must consider the possibility, however, that this information is connected in a global network. In my opinion, transmission of information in living systems by chemical substances such as DNA is not the whole truth. I see them as carriers of information, not the source of it. In recent years, physicists like Popp mentioned before developed hypotheses, according to which there is a constant flow of a coherent stream of a comparatively low number of photons within living organisms and probably also among them in close associations such as symbiosis. These photons are responsible for information transmission. DNA is not the only substance that can store information carried on by photons, though it is most efficient by far. One of the other substances with comparable abilities seems to be melanine, that is very widespread in fungi and animals, but also in soils. Organic soil with its high content of substances like melanine and microorganisms may proove a reservoir of stored information, that may be carried on by coherent, laser-like streams of photons under certain circumstances. There may be other substances keeping up the proposed network. Photons, moreover, mainly arrive on Earth from sun-light, so that even cosmic information may play a part. If something like a whole, Gaia is a holon again open to influences from the surrounding universe.

There is also a very interesting new hypothesis proposed by Markos (1995), which states, that the global information network necessary for the maintainance and development of the super organism Gaia might be upheld by prokaryotic microorganisms through their ability of almost unlimited gene transfer among individuals of different species (if the term species can be applied in their case at all). The reader is referred to his paper for further details. The author follows several trends of thought adopted also in this paper.

I ask you to take all this as mere suggestions to think about. They may open doors, but they also may lead to nothing at all. Nevertheless, I think it worthwhile to consider such possiblities.

To return to the main stream of our thought: we have a gradual decrease of potential information from the biosphere as a whole to living organisms with all the differences from more plastic to more specialized species, from them to organismic associations of every kind, of which symbiosis like that of lichens is nearest to true organisms, commensalism farthest, and finally to highly deterministic interrelationships like those between organisms and abiotic environment, among which open stands of vegetation under extreme conditions are not in need of any higher degree of flexibility resulting in a higher information content. Thus, organisms and associations related to them form a broad spectrum of decreasing potential information. Information content may be a measure for ecological amplitude in organisms and for similarity of any other system to true organisms. The biosphere (or Gaia?) may be considered a superorganism, but our lack of actual knowledge renders it impossible for the time being to give an unambiguous answer to that question, which is also crucial for the future of darwinism.

Literature

Clements, F.E. (1901): The Fundamental Principles of Vegetation. Amer. Ass. Advance. Science. Denver.

Clements, F.E. (1905): Research Methods in Ecology. (Univ. Publ. Co.) Lincoln, Nebrasca.

Clements, F.E. (1916): Plant Succession: Analysis of the Development of Vegetation. Publ. Carnegie Inst., Washington **242**: 1 – 512.

Koestler, A. (1967): The Ghost in the Machine. (Hutchinson) London.

Koestler, A. (1978): Janus. A. Summing Up. (Hutchinson) London.

Lovelock, J. (1979): Gaia: A New Look at Life on Earth. (Oxford Univ. Press) Oxford.

Lovelock, J. (1988):The Ages of Gaia. A Biography of Our Living Earth. (W.W. Norton & Co.) New York, London.

Markos, A. (1995): The Ontogeny of Gaia: The Role of Microrganisms in Planetary Information Network. J. theor. Biol. **176**: 175 – 180.

Odum, E.P. (1973): Fundamentals of Ecology. 3rd ed. (W.B. Saunders Co.) Philadelphia, Pa.

Phillips, J. (1935): Succession, Development, the Climax, and the Complex Organism: An Analysis of Concepts. Part II and III. J. Ecol. **23**: 210 – 246, 488 – 508.

Popp F.-A. (1986): On the Coherence of Ultraweak Photon Emission from Living Tissues. In: Kilmsten, C. W. (ed.): Disequilibrium and Self-Organisation. (D. Reviel) Dordrecht, Boston: 207 – 230.

Riedl, H. (in prep.): Living Organisms as "Elementary Particles".

Starr, M.P. (1975): A Generalized Scheme for Classifying Organismic Associations. In: Jennings, D.H., and Lee, D.L. (ed.): Symbiosis. Symposia of the Society for Experimental Biology XXIX. (Cambridge University Press) Cambridge, London, New York, Melbourne: 1 – 20.

Suess, E. (1875): Die Entstehung der Alpen. (W. Braumüller) Wien.

Tansley, A.G. (1935): The Use and Abuse of Vegetational Concepts and Terms. Ecology **16**: 284 – 307.

How to advance from the theory of natural selection towards the General Theory of Self-Organization

SIEVERT LORENZEN

How many theories of evolution do we have? One central or various competing ones? If the latter, are they compatible or incompatible with each other? If all biology is evolutionary biology, why is the validity of most published biological results not affected by the validity of any theory of evolution? Questions like these are still debated controversially or are simply neglected.

It is not my aim here to discuss the state and the validity of the theories of evolution debated or the reasons for their too insignificant impact on biological research. Instead, my aim is to identify the full extent to which the principle of natural selection is valid and to adjust its most general understanding to an adequate theory. The main results achieved are the following:

1) By discovering the principle of natural selection, Darwin discovered a novel natural law. It applies not only to organisms but, rather, to all units of the biotic and abiotic world which are able to reproduce themselves.

2) By appraising its full range of validity, the principle of natural selection is transformed into the natural law of self-organization, and the theory of natural selection into the General Theory of Self-Organization. This theory is of fundamental importance to the appropriate study of any kind of self-organization in both the biotic and abiotic world.

Darwin's suggestion: The principle of natural selection as the natural law of evolution

Obviously, all processes obey certain natural laws. This applies to evolution as well. Charles Darwin was the first to discover the natural law which, according to his theory of natural selection, is responsible for evolution. He called this law **the principle of natural selection** which emerged to be something like **the natural law of evolution**, because no further natural law was presented to be responsible for evolution.

The principle of natural selection is a law because it is valid without any exception and cannot be reduced to a more fundamental law. As such, it

- describes the conditions which make natural selection inevitable within all populations of organisms throughout all times and areas and

- describes the essential kind of results which inevitably are caused under these conditions.

These results are extremely diverse because of the many factors and feedbacks involved in the underlying processes. These evolutionary results may be explained by specific evolutionary theories.

The impact of the principle of natural selection on organisms and evolution shall be described in more detail:

Most important, organisms are able to reproduce themselves. By doing so, they generate and enhance growth pressures within their populations of conspecifics. All organisms need a certain amount of matter, energy and further resources in order to develop, sustain life and reproduce themselves. As the availability of at least some of these resources is limited per time and space unit, populations cannot grow infinitely. Instead, decreasing availability of resources leads to increasing inhibition of further population growth. Apart from this well known and important antagonist of population growth, there are further antagonists which are also evoked and enhanced by the growing populations themselves. They include accumulation of wastes, increasing exploitation by grazers, predators and parasites, increasing danger of contagious diseases, and increasing resistance of food organisms against growing populations of their exploiters.

Other antagonists of population growth such as climatic conditions do not depend on particular population densities.

The stronger the antagonists of population growth are, the more reproductive success those organisms will have which succeed in

1) getting access to sufficient resources, even if the latter are scarce,
2) taking best advantage of even small amounts of resources,
3) coping best with all other difficulties they meet in their particular environments,
4) invading new areas in which the resources needed are less scarce or less contested,
5) altering their ontogenesis in response to altering environmental conditions, or
6) exploiting considerable amounts of novel resources.

According to Darwin's terminology, organisms which are among the most able in making use of one or several of these possibilities are called **the fittest**. According to Darwin's theory of natural selection, the fittest will have more reproductive success than their less fit conspecifics. Therefore, in the course of many generations, the fittest organisms will contribute more to the cohesion of their populations than their less fit conspecifics. As a result, populations become recognizable as coherent dynamic structures which are known as species, subspecies, varieties or something else. By invading new areas, organisms may promote spatial dispersion of their populations, and by exploiting new resources, they may give rise to sympatric species cleavages.

In any case, antagonists of population growth must be weak enough so as not to inhibit population growth at all.

Fitness differences between conspecifics may be generated by various mechanisms, for instance by

– mutations of genes,
– all kinds of recombination of alleles,
– horizontal gene transfers,

– symbiotic associations of heterospecific organisms, and
– alterations of environmental conditions.

The latter mechanism deserves special attention, as it points to a twofold impact which environments may exert on organisms: 1) The fitness of any organism is determined not only by its own constitution but by that of its environment as well. Therefore, organisms with low local fitness are not necessarily condemned to early death but, instead, may alter their fitness by getting access to altered sets of environmental conditions. Although many of these efforts will decrease the fitness of organisms, others will increase it and, hence, will be of evolutionary importance. 2) Environmental conditions may exert influence on the phenotypic development of organisms and, hence, on their fitness.

Evidently, the concept of fitness does not refer to success achieved but, instead, to the ability to achieve success. If fitness were equalled to success, the notion "survival of the fittest" would degenerate into a mere tautology (for a thorough analysis of this problem see Waddington 1975, chapter 25). Nevertheless, such tautology seems to be suggested commonly by population and evolutionary geneticists, for example by Maynard Smith (1989: 36) who suggests taking "the survival and reproduction of the different genotypes" to be the "measure of fitness". Apparently, what is meant seems to be the following: Any success achieved by a particular type of organisms exposed to a particular set of environmental conditions will be exemplary in predicting, that other organisms of the same type will achieve the same success if exposed to the same set of environmental conditions.

This is the current understanding of the principle of natural selection.

Transformation of the principle of natural selection into the natural law of self-organization, and of the theory of natural selection into the General Theory of Self-Organization

According to a strict philosophical demand, the validity of any natural law must be independent of particular times, particular localities, and particular objects. As the principle of natural selection is acknowledged as a natural law, it must meet the three criteria listed as well. In the past, the validity of the first two criteria was always acknowledged correctly, whereas that of the third was generally neglected. This means, no answer was given to the question of what exactly is the complete set of objects to which the principle of natural selection applies.

Apparently, organisms are among these objects. Further objects were discovered within the last decades; they belong to each the biotic and abiotic world and include molecules of RNA within energy rich liquids (Eigen & Schuster 1977) and light waves within energy rich LASER-systems (Haken, e.g. 1983). Despite the novel results, the question raised was not answered.

Recently, while writing a chapter on the theory of evolution for a forthcoming handbook on environmental research, I became aware of this question and found the following answer to it (Lorenzen 1997a):

The essential properties shared by exactly all objects to which the principle of natural selection applies, are the following:

a) These objects are capable of reproducing themselves, regardless of whether this occurs by division, autocatalysis, autinduction, by sexual reproduction or other means.

b) For reproduction, these objects need sufficient amounts of suitable resources.

c) Fitness differences influence the reproductive success of these objects.

d) Fitness differences are generated again and again within all populations of these objects.

e) These objects are mortal.

These objects will be termed **reproductive units**. They include organisms, cells, DNA molecules, reaction cycles, hypercycles, autocatalytic substances, fashions, optic and acoustic waves, and so on, whenever they occur in appropriate, energy rich systems suitable for their reproduction. In many cases, reproduction of reproductive units cannot be observed directly but, rather, is perceived as self-enhancement of a dynamic phenomenon. In all cases, reproduction causes metabolism of the respective populations.

As the resources needed are always limited per time and space unit, the principle of natural selection governs reproduction of all reproductive units in exactly the same way as it was described above for organisms. If sufficient amounts of suitable resources are available and if population growth is inhibited neither too weakly nor too strongly, dynamic order may emerge as a result. In all cases, dynamic order disappears, when supply with resources leaves off or when other inhibitors of reproduction become too strong.

Organisms themselves are composed of many populations of reproductive units. These include cells, DNA molecules and various kinds of reaction cycles and hypercycles. Additionally, organisms are reproductive units within their respective populations of conspecifics. Generally, any coherent population of reproductive units may itself be a reproductive unit at a higher level.

An inspection of contemporary literature reveals that the principle of natural selection as outlined governs exactly all processes which are generally subsumed under **self-organization** of persistent dynamic structures. Despite a large amount of literature on this topic (e.g. Haken 1983, Prigogine & Stengers 1986, Probst 1987, Krohn & Küppers 1992, Kauffman 1993), the concept of self-organization was not defined unequivocally and a central theory of self-organization was not suggested. It is for this reason that the following suggestions were made (Lorenzen 1997a):

–   With regard to its whole range of validity, the principle of natural selection deserves a new name. **The principle of self-organization** has been suggested. In order to stress its character as a natural law, it will now be termed **the natural law of self-organization**. It cannot be reduced to some more fundamental law.

–   The concept of **self-organization** has been suggested to refer exclusively to dynamic processes which are governed by the natural law of self-organization.

–   The transformation of the principle of natural selection into the natural law of self-organization requires an extension of the theory of natural selection. This extension has been called **the General Theory of Self-Organization** which refers exactly to all reproductive units which share the properties listed above under a - e.

The General Theory of Self-Organization reads as follows: Whenever reproductive units (objects which fulfill the properties listed above under a - e) have access to sufficient amounts of resources needed for their reproduction, the latter is governed by the natural law of self-organization. That is, in the course of many generations, any population of such reproductive units will become recognizable as a structure which may be either dynamically coherent or dynamically chaotic.

As the General Theory of Self-Organization refers to a natural law and its impact on all reproductive units, it cannot be reduced to some more fundamental theory and, hence, is itself a fundamental theory. As such, it does not explain specific cases of self-organization but, instead, focusses on the essential aspects which must be studied in detail in order to understand any specific case of self-organization. Even when phenomena are highly complicated, underlying laws may be very simple.

All theories which deal with specific cases of self-organization may be called **specific theories of self-organization**. Among them is the Synthetic Theory of Evolution which illuminates the impact of hereditary and environmental events on both, organisms and evolution.

Nonlinearity, self-organization, and chaos science

The natural law of self-organization includes the most general formulation of any **nonlinear law** governing any kind of self-organization: Reproduction causes both positive and negative feedback in population growth. Under favourable conditions, positive feedbacks are responsible for exponential population growth; however, growing populations induce and enhance antagonistic reactions as well which serve as negative feedbacks on population growth. Any combination of at least one positive with at least one negative feedback is a nonlinear growth law. Sciences of chaos and self-organization deal precisely and exclusively with these laws which, normally, are written down as nonlinear differential or difference equations.

It should be mentioned that there are nonlinear laws which are irrelevant to the self-organization of persistent dynamic structures. Any combination of only positive or only negative feedbacks refers to such laws.

The science which deals with self-organization as outlined was called **synergetics** by Haken in 1969 (see Haken 1983). This term will be adopted herein. In synergetics, special emphasis is laid on the study of those parameters of nonlinear growth laws, which allow populations of reproductive units to develop into structures of persistent dynamic order or persistent dynamic chaos.

In addition to the principle of natural selection, Haken introduced the novel **slaving principle** and the so-called **order parameter** and stressed their essential importance to any kind of self-organization of dynamic structures. This suggestion is not included in the General Theory of Self-Organization. Why?

At first glance, the slaving principle conveys the impression that it refers to the impact of intraspecific interactions on the coherence of populations. However, a closer analysis reveals that all effects evoked by the slaving principle are identical with those evoked by the natural law of self-organization. Therefore, the slaving principle is synonymous with the natural law of self-organization (Lorenzen 1997a and b).

The order parameter refers to nothing else than the average fitness of conspecifics within a local population of reproductive units (Lorenzen 1997b). Again, this aspect is covered completely by the natural law of self-organization.

Chaos science deals with dynamic chaos and dynamic order resulting from processes governed by nonlinear growth laws. These laws do not differ from those studied in synergetics, as they deal with antagonistic effects of positive and negative feedbacks on the development and maintenance of dynamic structures. Fitness differences between reproductive units are largely neglected in chaos science, although they occur: The more a critical population density is surpassed, the lower the fitness of the respective reproductive units is. It is for this reason that synergetics and chaos science are identical.

In his book "The origins of order – self-organization and selection in evolution", Kauffman (1993) made self-organization *and* selection responsible for evolution. Any inclusion of the concept of self-organization in the study of evolution deserves appreciation. However, Kauffman failed to extend the theory of natural selection to a more fundamental theory, because he failed to extend the principle of natural selection to a more fundamental law; specifically, he did not notice that, at all levels, self-organization without selection is impossible. The concept of self-organization embraces the concept of selection. By overlooking this essential aspect, Kauffman failed to discover the natural law of self-organization.

Results from synergetics have provided convincing evidence that any process governed by the natural law of self-organization may produce the following two effects which are often not perceived by intuition:

– Depending on both the numerical value of parameters involved in a specific law of self-organization and the initial conditions, self-organization may result either in different kinds of dynamic order or in different kinds of dynamic chaos. That is, dynamic order and dynamic chaos are the two sides of one coin. In its state of dynamic order, any population will be called a **coherent dynamic structure** (Lorenzen 1997a), and in its state of dynamic chaos, any population will be called a **chaotic dynamic structure**.

– By self-organization, any population of reproductive units may be highly resistant to even severe perturbations on the one hand but, at critical states, extremely sensitive to even slightest perturbations on the other hand. In cases like the latter, dynamic order may rapidly emerge from dynamic chaos or vice versa, or an existing dynamic order may rapidly be replaced by another dynamic order. The two kinds of response to perturbations are unknown from processes governed by linear laws.

For centuries, natural science has focused on processes which may be analysed without referring to the natural law of self-organization. Although such science allows a large variety of predictions to be made, it is unsuitable for studying or understanding essential aspects of self-organization, evolution included. This may be the fundamental reason why the theory of natural selection and its implications were so difficult to understand for such a long time and are so even now.

Impact of the General Theory of Self-Organization on biological research – the art of conclusion by analogy

All kinds of biotic self-organization may affect the fitness of organisms and, hence, play a role in evolution, i.e. in the long-term self-organization of all biotic phenomena we know. This aspect may be expressed by a theory to be inferred from the General Theory of Self-Organization and which may be called the **General Theory of Evolution**. According to this theory, origin and maintenance of any dynamic biotic structure is brought about by the natural law of self-organization. That is, in any of these cases, the conditions listed above under a-e are fulfilled and may be explored in detail by employing the concept of self-organization. This type of work may be illustrated by the following three examples:

1) Conclusions by analogy in studying phylogenesis and ontogenesis. Recently, at the 38th German Phylogenetic Symposium in Giessen/Germany in 1996, the claim was made that the laws of phylogenesis and ontogenesis are completely different, because phylogenesis advances by genetic changes and natural selection, whereas any ontogenesis is the realization of a specific genetic program. Therefore, novelties might arise in phylogenesis but not in ontogeneses. This claim includes the notions that the natural law of self-organization governs phylogenesis but not ontogenesis and that genomes are something like creators of organisms.

These claims must be rejected, because their underlying argumentation contradicts a valid and very powerful way of interdisciplinary reasoning which consists in **conclusions by analogy**. Analogies of this type are not based on superficial similarities between different objects but, instead, refer to a strictly controlled transfer of knowledge from one system to another: If different systems share equivalent conditions, they may respond equivalently to the same natural laws. Therefore, any result achieved in the one system corresponds to an equivalent result in the other and vice versa. Conclusions by analogy precisely make use of these equivalences.

All prerequisites of employing this kind of argumentation are completely fulfilled in the cases of phylogenesis and ontogenesis: In both cases, there are reproductive units like organisms, cells, DNA, etc which share the properties listed above under a-e, and there are sufficient amounts of resources available for their reproduction. Therefore, the natural law of self-organization applies to all of these reproductive units. Particularly, it is wrong to assume that phylogenesis and all kinds of ontogenesis might be governed by different laws. On the contrary, any specific theory of self-organization which is valid within a phylogenetic context is equivalently valid within an ontogenetic context and vice versa. Genotypes cannot act as creators of organisms but, rather, act as specific sets of conditions involved in the self-organization of organisms.

Conclusions by analogy are rather common in mathematics and may be exercised in many other sciences as well. Examples are provided by Meinhardt (1995) in his book on biotic pattern formations. This book demonstrates the wide range of dynamic patterns which may be generated by simply modifying some parameters of a small set of nonlinear differential equations. Another example was provided by Edelman (1993) who suggested that both, phylogenesis and neural self-organization, are governed by the principle of natural selection in precisely the same way. To underline this view, he introduced the term "neural Darwinism".

2) Diversity of evoked inhibitors of population growth. In the course of various ecosystem projects, important results referred to the discovery of unconventional inhibitors of population growth induced and enhanced by the growing populations themselves. Two examples may serve for illustration: 1) By their exudates, aquatic predators like larvae of the midge *Chaoborus*, notonectid water bugs and various fish species induce the formation of particular phenotypes or particular kinds of behaviour in their respective prey populations of certain species of *Daphnia* (Cladocera), thereby providing the latter with increasing protection against attacks by their specific predators (Jacobs 1987). 2) By their feeding on leaves of the larch *Larix decidua*, growing populations of the larch bud moth *Zeiraphera diniana* induce the formation of modified leaves which are unsuitable as food to the larvae of larch bud moths and, hence, cause the collapse of the larch moth populations (for a recent review see Baltensweiler 1996).

3) Saltations in evolution. The occurrence of saltations in evolution is commonly strongly denied, even by the punctualists (Gould & Eldredge 1986). According to results from synergetics, the emergence (i.e. the sudden origin) of novel coherent dynamic structures is a rather common and familiar event to be observed in any kind of self-organization. Examples of saltations are known even from present-day species: Whenever different life forms are realized within a species, the transition from one life form to another may occur by saltation, i.e. without the formation of intermediate stages. Saltations do not need to be results of speciation events but, rather, may occur within species.

One striking example of saltations refers to species of the nematode genus *Deladenus* (see Bedding 1968): The free-living specimens look like members of the family Neotylenchidae, whereas insect parasitic specimens look like members of the very different family Allantonematidae. Each life form may be realized during innumerable generations. There are other genera of the Neotylenchidae and Allantonematidae, in which the species develop only one life form, either the free-living or the insect parasitic. Intermediate stages between the neotylenchid and allantonematid habitus are unknown. Therefore, at least a part of allantonematid species must have evolved from neotylenchid-like ancestors by decoupling the insect parasitic life form from the free-living. Any sudden origin of a novel life form may be subsumed under saltation.

Generally, a deep understanding of a large variety of biotic processes, evolution included, may be expected by employing the General Theory of Self-Organization as a powerful tool of analysis.

Literature

Baltensweiler, W. (1996): Kurzzeit- versus Langzeitstudien – Resultate oder Erkenntnisse? Erfahrungen aus der Forschung über den Lärchenwickler. Verh. dt. zool. Ges. **89**,2: 181–188.

Bedding, R.A. (1968): *Deladenus wilsoni* n.sp. and *D. siricidicola* n.sp. (Neotylenchidae), entomophagous-mycetophagous nematodes parasitic in siricid woodwasps. Nematologica **14**: 515–525.

Edelman, G.M. (1993): Neural Darwinism: Selection and reentrant signaling in higher brain function. Neuron **10**: 115–125.

Eigen, M. & Schuster, P. (1977): The hypercycle. A principle of natural self-organization. Part A: Emergence of the hypercycle. Naturwissenschaften **64**: 541–565.

Gould, S.J. & Eldredge, N. (1986) Punctuated equilibrium at the third stage. Syst. Zool. **35** 143–148.

Haken, H. (1983): Synergetics. An introduction. 3rd edition. Springer, Berlin.

Jacobs, J. (1987): Cyclomorphosis in *Daphnia*. In: R.H. Peters & R. de Bernardi (Eds): *Daphnia*. Mem. Ist. ital. Idrobiol. **45**, pp 325–352.

Kauffman, S.A. (1993): The origins of order. Self-organization and selection in evolution. Oxford Univ. Press, New York, Oxford.

Krohn, W. & Küppers, G. (Eds) (1992): Emergenz: Die Entstehung dvon Ordnung, Organisation und Bedeutung. Suhrkamp, Frankfurt/M.

Lorenzen, S. (1997a): Das Selektionsprinzip universell gültig als Naturgesetz – von der Evolutionstheorie zur Allgemeinen Selbstorganisationstheorie. In: O. Fränzle, F. Müller & W. Schröder (Eds): Handbuch der Umweltwissenschaften. Grundlagen und Anwendungen der Ökosystemforschung. Ecomed, Landsberg. Kapitel III-2,2: 14 pp.

Lorenzen, S. (1997b): Einführung der Allgemeinen Selbstorganisationstheorie als Haupttheorie der Synergetic. Ethik und Sozialwissenschaften 7: 622–624.

Maynard Smith, J. (1989): Evolutionary genetics. Oxford Univ. Press, Oxford, New York, Tokyo.

Meinhardt, H. (1995): The algorithmic beauty of sea shells. Springer, Berlin, Heidelberg, New York.

Prigogine, I. & Stengers, I. (1986): Dialog mit der Natur. Neue Wege naturwissenschaftlichen Denkens. 5th edition, Piper, München, Zürich.

Probst, G.J.B. (1987): Selbst-Organisation. Ordnungsprozesse in sozialen Systemen aus ganzheitlicher Sicht. Parey. Berlin, Hamburg.

Waddington, C.H. (1975): The evolution of an evolutionist. Edinburgh University Press.

The evolutionary periodicity of flight

ANTONIO LIMA-DE-FARIA

1. Introduction

Chemical periodicity took over a century to elucidate. Although information was available since 1772 on the regularity of certain properties of the chemical elements, no one could think of any system of order among the elements for 100 years.

In 1865 Newlands proposed his law of octaves, showing that the properties of certain elements repeated themselves and that they could be put together in groups of eight. He sent his paper to Nature, which was rejected and ridiculed.

Six years passed, and in 1871, Mendeleev produced periodic tables in which he left empty spaces that represented elements not yet discovered, but which he could predict their properties. His predictions about these properties were confirmed a few years later when new elements were discovered, and the concept of periodicity soon became established in chemistry.

However, no chemical or physical mechanism was available that could explain this ordered system of properties. It took 50 years more until this periodic system was clarified. Only when the electronic organization of the atom became established in the 1920s and 1930s did it become evident that it was, not the total number of electrons, but only the number of electrons that occur in the outer shell of an atom that governs its chemical properties.

A most significant feature of chemical periodicity is that chemical elements exhibit similar properties, irrespective of their degree of complexity. A simple atom such as beryllium (Be, atomic mass 9.0) will have the same properties as a most complex one, such as radium (Ra, atomic mass 226.0). Both are alkaline earth metals which are silvery white, lustrous and relatively soft. The same is true for helium (He, atomic mass 4.0) and radon (Rn, atomic mass 220) which are both inert gases. This means that extreme differences in complexity of organization among chemical elements do not have an influence on the same basic properties of these elements. Moreover, not just one, but a package of as many as 10 properties, are repeated in the elements that happen to have the same electron number in their outer shell (Fig. 1).

Although chemical periodicity is now firmly established and has become the basis of the interpretation for most chemical phenomena, there are irregularities. Not less than 700 different periodic tables have been published and these mirror the presence of exceptions and of properties that are still difficult to explain. The location of helium (He) in the tables varies with the author; moreover, copper (Cu) and chromium (Cr) do not fit easily into the periodicity charts. The length of the periods built by the elements is not constant either. Although there is some variation in their properties, periodicity of the elements has become a most powerful instrument in the understanding of chemical transformations (Fig. 2, Sanderson, 1967; Greenwood and Earnshaw, 1989).

2. From Chemical To Biological Periodicity

At present the concept of periodicity may seem as inappropriate to biologists as it was to chemists in the first half of the nineteenth century. However, the advancement of molecular biology has been so rapid over the past decades that we have been obliged to revise or abandon some of our previous concepts.

The search for periodicity among living organisms discloses that, as in the case of the chemical elements, not just one property, but a series of biological functions and structures display periodicity. Moreover, periodicity results, as in the case of chemical elements, in the emergence of the same property in the simplest as well as the most complex organisms. Likewise, a small versus a large amount of DNA or the presence of few genes versus a vast array of them are not limiting factors to the periodic emergence of the same basic structures and functions. Also, as in the case of the atoms, the length of the period does not need to be always the same.

Periodicity provides an explanation of the sudden emergence of features that had been absent in species for millions of years of evolution. An example is the placenta. This organ appears at first in some of the most primitive invertebrates, such as the Onychophorans and then turns up in fishes only to disappear once again. Later the placenta is found in other groups of vertebrates, such as amphibians and reptiles, but it again becomes absent in the marsupial mammals and suddenly reappears as the dominant feature of the placental mammals. In all these different organisms the function of the placenta is essentially the same and it attains a high degree of complexity in all forms possessing it. Besides the placenta, other functions display periodicity. Some of them may be mentioned such as the recurrent emergence of the penis, the repeated appearance of bioluminescence, the punctuated form of advanced vision in invertebrates and the capacity to fly (Lima-de-Faria, 1994, 1995, 1997). In this article I will concentrate on the periodicity of flight (Figs. 3 and 4).

3. The Periodicity Of Flight

The Chordates, which include the vertebrates, constitute a single phylum. But the invertebrates have diversified into 30 phyla. Yet in only one of the invertebrate phyla, the Arthropoda, did flight emerge, and within this phylum only the class Insecta developed this ability. No other invertebrate group had shown this phenomenon before the insects, and no other invertebrate group displayed it in the future. This was not due to lack of time, since at least 600 million years have elapsed, since the invertebrates emerged. Such a punctuated appearance of this function is one of the characteristic features of biological periodicity. No other ancestral groups possess wings and no other invertebrate phyla that subsequently evolved were able to fly. An additional characteristic is that flight at its origin (350 to 270 million years ago) it had already attained full capacity. Fossils of the giant dragonfly *Meganeura*, which inhabited the forests of the Carboniferous period, had wings that measured 73 cm from tip to tip. Many bird species have smaller wings and present-day insects have wings of smaller size (Fig. 5).

The second time flight appeared it showed no direct connection with that of insects. As the vertebrates emerged, the fishes, the amphibians and most reptiles could not fly. The flying reptiles, the pterosaurs (230 to 195 million years ago) had wings supported by a different type of skeleton from that in insects, but which were equally efficient in allowing the pterosaurs to fly.

The pterosaurs became extinct during the Cretaceous period leaving no survivors (Walker, 1974). Later, the situation repeated itself. The birds, which emerged 180 to 135 million years ago, had as ancestors flightless reptiles such as the Thecodonts which were related to the dinosaurs, and were the main reptile group during the Triassic period (Whitten and Brooks, 1988).

The evolution of mammals started about 100 million years ago, with the marsupials appearing first around 63 million years ago and the placental orders about 58 million years ago. During their long evolution the mammals diversified into over 20 orders, without any indication of flying capacity. Within this large number of flightless families and species, appeared a single order that has no direct connection with the birds or the flying reptiles, but which evolved highly effective wings. These are the bats which emerged about 40 million years ago. Fossils that represent intermediate stages between the ancestors of bats and fully evolved bats have so far not been found; the earliest existing fossil bats are so well preserved that they disclose the stomach contents of these first bats (Colbert, 1980).

Equally showing no signs of direct relationship to other groups of flying vertebrates was the emergence of flight among fishes. These vertebrates arose about 425 million years ago and are represented today by a large number of species. The teleosts appeared 190 million years ago. Among them only a few species can fly, and do so in an environment adverse to their respiration, namely in air. These flying fish are obliged to return within minutes to water otherwise they would die, but their flight shows a degree of complexity considered not inferior to that of insects (Hanström and Johnels, 1962; Beazley, 1980).

The evolution of flight discloses that in non-related groups and at unexpected occasions, animals that did not fly gave rise to species that became aerial with high efficiency. Most significant for the understanding of periodicity, is the fact, that as in the chemical elements, the degree of complexity of the organism is not a limiting factor to the emergence of the periodic features, simple invertebrates like the insects, as well as the complex mammals, such as the bats, display the same high capacity to fly (Figs. 6 and 7).

Several bird and insect species have wings but do not fly. A complex of parts, organs and specific molecules must be present for flight to occur. The most important are: (1) The animal skeleton, which may consist of bones or other types of tissue which have a supporting function; (2) Articulations, involved in the change of position of the supporting elements; (3) Powerful muscles that control the various movements; (4) Additional energy, that is used during flight; (5) Great quantities of blood and air, that are necessary during the rapid wing movements; and (6) An efficient nervous system, that is able to control direction, stability and speed of flight.

The comparison between insects, pterosaurs, birds, fishes and bats reveals that similar basic solutions for flight were used in these different groups. A coherent and rigid pattern recurred in all of them, namely: (1) A fusiform body is present in most groups; (2) The wings are located above the center of gravity on the dorsal surface of the body in both invertebrates and vertebrates; (3) Air filled bones and other organs are present in several groups; (4) Antagonistic muscles and articulations are obligatory components; (5) The same source of energy, glycogen, is used by insects and birds; and (6) Measurement of speed, balance and vision are controlled by the nervous systems of the different organisms, although the brain of an insect is at least one hundred times smaller than that of a bat (McFarland, 1981).

Flight is a phenomenon distinct from gliding. Fishes do not perform a passive gliding as some squirrel species do, but use powerful muscular contractions that oblige the body and the wings to vibrate like those of insects (as much as 70 contractions per second). Fishes can circle around a boat, can stop in air, and modify their flight direction at ease, as birds do (Hanström and Johnels, 1962).

Another particular feature is the skeleton of insects, that does not consist of bones and is thus different from that of vertebrates, yet allows the same solution for flight. A corresponding situation occurs in vertebrates. They have evolved organs that do not either possess a bony skeleton but are equally functional. An example is the large caudal fin of whales which can measure over a meter in height. This body organ did not exist in the ancestors of whales and has no skeleton. Yet it is hardly distinguishable, in its form and function, from the caudal fin of a fish which has a well developed bony skeleton. Hence, both in mammals and insects, organs that are practically identical in form and function can be produced with or without bones (Macdonald, 1984).

The periodicity of flight is characterized by the following features: (1) Flight arises without immediate predecessors in five major groups; (2) Few intermediate forms were involved or it appeared suddenly; (3) It tends to occur in a restricted class or order; (4) It consists of a coherent package of structures and functions, several of which being obligatory components in all the groups; (5) The evolutionary position of the organism is not directly related to the complexity of the phenomenon as the flight of bats is no more complex than that of birds; (6) The number of wings is not either correlated to the evolutionary position of the species in that birds have only two wings, whereas insects and fishes may have four; (7) There seems to be no obligatory relationship to a specific type of environment, since fishes fly in a medium (air) adverse to their normal respiration, which takes place in water. Moreover, all mammal orders, except the bats, remained terrestrial, and among all the aquatic and terrestrial invertebrates, only, the insects conquered the air; and (8) The interval at which flight has occurred has been 80 to 110 million years which is rather regular. Yet, it is expected that the length of the period may not be always the same because this has already been demonstrated to be the case with the chemical elements (Lima-de-Faria 1994, 1995, 1997).

4. Molecular Mechanisms That May Be Responsible For Periodicity

When a phenomenon is recognized, it can not be expected that the mechanism responsible for its occurrence becomes immediately understood. However, molecular biology has developed so rapidly, in the last two decades, that several main mechanisms for biological periodicity can be considered.

Most structural genes in multicellular organisms consist of intervening DNA sequences termed introns and other sequences coding for proteins termed exons. These two types of sequences alternate and can recombine. New mosaic proteins are formed when exons located originally with other DNA sequences move to new locations. Examples are the cholesterol-transport protein and the human tissue-type plasminogen activator (Südhof et al., 1985a and b; Ny et al., 1984). Still more interesting is alternative RNA splicing, since it does not change the DNA sequences of the chromosomes. Yet the mechanism modifies the messenger RNA with the result being the formation of different proteins. The phenomenon is reversible. This is most significant to understand periodicity as: (1) It is not necessary to change the genetic constitution to produce a different protein; and (2) an old protein can suddenly reappear due to the reversibility of the RNA splicing. Examples are the sarcomeric contractile protein genes. This means that new and old proteins can be formed without changing appreciably the genetic constitution of the organism (Breitbart et al., 1987; Smith et al., 1989; Van der Ploeg, 1990).

The emergence of coherent packages of structures and functions which are involved in flight can be better understood on the light of internal cascades of molecules that have been studied recently. The coherence arises from the fact that the initial production of a single molecule obliges the release of a series of other substances that follow in a cascade and which have simultaneously morphogenetic effects on a group of different tissues and organs.

Blood clotting in humans is one example of a molecular cascade. It involves more than 10 proteins that follow in a rigid sequence determined by intrinsic molecular events (Ryden and Hunt, 1993). Another case with special significance for periodicity is the action of the hormone produced by the anterior pituitary, an endocrine gland located in the brain. The action of a single molecule results in radical functional changes in quite different body organs. As a result of three successive cleavages a single polypeptide is carved into six hormones. The resulting amino acid sequences: (1) act on the adrenal cortex; (2) affect lipid metabolism; (3) change the pigment of the skin; and (4) modify brain function (Fig. 8). Thyroxine in frog embryos has similar effects. Its action leads to drastic changes in: (l) the heart; (2) the digestive apparatus; and (3) the muscles of the larva. Moreover, new organs are built which lead to the emergence of the adult. These are, among others, the lungs and a complete vertebrate skeleton with adjoining muscles forming body members not present in the larva. Controls show that in the absence of thyroxine no organ transformations occur (Alberts et al., 1994).

The most pertinent example is the molecular cascade that results in the formation of the placenta, which is an organ that also shows periodicity. This cascade has been studied in detail. The pathway, which is strictly canalized, starts with the follicle-stimulating hormone (FSH). This increases the cellular concentration of cyclic AMP. As a result proteins are produced that affect the transcription of DNA. A second cascade is started by the hormone estrogen and as a result other molecules are formed which act on the uterus and lead to the development of the placenta (Fig. 9; Dorrington, 1979).

The wing of an insect has been considered by zoologists to be analogous, to that of a bird, but not homologous. The reason was that the two structures are so different that they seemed to have no phylogenetic relationship. The recent work on embryonic development in *Drosophila* and vertebrates (including birds and mammals) has disclosed that the same type of genes are responsible for the formation of corresponding regions of the body in invertebrates and vertebrates. These genes, called homeotic genes, not only control the formation of similar structures but have also retained the same arrangement in the chromosomes of the distantly related species. The wings of birds are mainly formed by the bones and muscles of the anterior limbs. It has been found by molecular analysis, that the genes responsible for limb formation in the chicken are homologous to the genes that lead to the production of the wings in *Drosophila*. The results show that the wing of an insect is determined by the same type of genes that decide the formation of the wing of a chicken. The analogous relationship of these features has become a homology since it is the product of the ordered expression of the same DNA sequences (Affolter et al. 1990, De Robertis et al. 1990, Holland 1992, Lawrence 1992, McGinnis and Kuziora 1994).

Chemical periodicity has turned out to be a most fruitful finding that allows the explanation of many of the phenomena encountered today in chemistry. Many of the problems that remain to be elucidated in evolution may turn out to be more intelligible in the light of biological periodicity (Lima-de-Faria 1998a,b).

5. The Direct Production Of Flies With Four Wings

Confirmation that recombination between genes may produce suddenly a structure and function that has not been available for millions of years is furnished by the latest results in molecular development, studied in *Drosophila*. A normal fly has only two wings, but molecular biologists have recently been able to produce flies with four wings. What is remarkable is that these flies are not monsters but are quiet well formed and their new wings are equally normal. These accessory wings are also partly functional. Moreover, contrary to the generally accepted idea on the mechanism of evolution, these flies were neither obtained by selection of individuals that originally displayed rudimentary wings, nor by selection that led to successive progenies with bigger and better wings. The flies with four wings were obtained directly, without any progressive transformation.

The studies of Lewis (1978), Bender et al. (1983), Peifer et al. (1987), Duncan (1987) and Lawrence (1992), led to the production of four wing flies by a simple molecular manipulation combining mutations occurring in the genes *bithorax* and *postbithorax*. These genes were somatically recombined following irradiation of larvae the result being the immediate appearance in the progeny of normal flies with four perfect wings (Fig. 10 and 11).

Edward B. Lewis has supplied the following personal communication on November 11th 1998 (T3 is the segment of the thorax closest to the abdomen; normal wings are formed on T2 and the second pair on T3). "We are trying hard to get better muscle development in T3 by adding the trithorax mutant to the triple mutant bx genotype but it will be a month before the matings are completed to generate these flies. As I indicated the problem is that ectoderm and mesoderm are out of phase by one whole parasegment so Antpennapedia + function is presumably on in T3 mesoderm and preventing it from developing flight muscles. Trx mutants should given partial and variable loss of that function so we might get mosaic flies with complete T3 muscles developed. I have movies that show by the way that the second pair of wings do move up and down a little. I think this can be seen in the movie of the triple mutant fly that I showed in my Nobel lecture."

Insects with four wings are known to exist in nature and to be quite common, an example is the butterflies. Moreover, the position of the wings in the butterflies is the same as in the four-winged *Drosophila*. Among the vertebrates fishes may display four wings as well.

The implications of this molecular transformation at the chromosome level are most significant for the understanding of evolution and of biological periodicity. They reveal the following. 1) The formation of new wings is an immediate and direct process. 2) The structures appear as "ready made" the pattern being fully coherent from the onset. 3) The structures are accompanied by the corresponding functions, such as blood flow, control by the nervous system, access to large amounts of energy. 4) The position of the wings is not accidental but occupies the dorsal part of the body. 5) The genetic transformation apparently reenacts in days what has occurred millions of years ago.

Literature

Affolter, M., Schier, A. and Gehring, W.J. (1990): Homeodomain proteins and the regulation of gene expression. Curr. Opin. Cell Biol. **2**: 485–495.

Alberts, B., Bray, D., Lewis, J., Raff, M., Roberts, K. and Watson, J.D. (1994): Molecular Biology of the Cell. Garland Publishing, New York, London.

Beazley, M. (1974): The World Atlas of Birds. Mitchell Beazley Publishers Limited, London.

Beazley, M. (1980): The Atlas of World Wildlife. Rand McNally and Company, printed in the Netherlands.

Bender, W., Akam, M., Karch, F., Beachy, P.A., Peifer, M., Spierer, P., Lewis, E.B. & Hogness, D.S. (1983): Molecular genetics of the bithorax complex in *Drosophila melanogaster*. Science **221**: 23–29.

Borror, D.J., DeLong, D.M. and Triplehorn, C.A. (1976): An Introduction to the Study of Insects. 4th Ed. Holt, Rinehart and Winston, New York, pp. 1–812.

Boule, M. and Piveteau, J. (1935): Les Fossiles. Elements de Paléontologie. Masson & Cie, Editeurs, Paris.

Breitbart, R.E., Andreadis, A. and Nadal-Ginard, B. (1987): Alternative splicing: A ubiquitous mechanism for the generation of multiple protein isoforms from single genes. Ann. Rev. Biochem. **56**: 467–495.

Brumpt, E. (1936): Précis de Parasitologie. II. Collection de Précis Medicaux, Masson & Cie, Paris.

Colbert, E.H. (1980): Evolution of the Vertebrates. A History of the Backboned Animals through Time. 3rd ed. A Wiley-Interscience Publication, John Wiley and Sons, New York, pp. 1–510.

De Duve, C. (1984): A Guided Tour of the Living Cell. Volume 2. Scientific American Library, Scientific American Books, Inc., New York.

De Robertis, E.M., Oliver, G., and Wright, C.V.E. (1990): Homeobox genes and the vertebrate body plan. Sci. Amer. **263** (1): 26–33.

Dorrington, J. (1979): Pituitary and placental hormones . In: "Mechanisms of Hormone Action, Reproduction in Mammals, Book 7" (Eds. C.R. Austin and R.V. Short), Cambridge University Press, Cambridge, pp . 53–80.

Duncan,I. (1987): The bithorax complex. Ann.Rev.Genet. **21**: 285–319.

Francé, R.H. (1943): La Maravillosa Vida de los Animales. Una Zoologia Para Todos. Editorial Labor, S.A., Barcelona.

Freeman, R. (1972): Classification of the Animal Kingdom. An Illustrated Guide. Hodder and Stoughton Ltd., The Reader's Digest Association Ltd., London, pp. 1–55.

Greenwood, N.N. and Earnshaw, A. (1989): Chemistry of the Elements. Pergamon Press, Oxford, New York, Beijing, pp. 1–1542.

Hanström, B. and Johnels, A.G. (1962): Benfiskar. In: "Djurens Värld, Band 6, Fiskar: 2" (Ed. D. Hanström), Forlagshuset. Norden AB, Malmö.

Holland, P. (1992): Homeobox genes in vertebrate evolution. Bioessays **14**: 267–273.

Lawrence, P.A. (1992): The Making of a fly. The Genetics of Animal Design. Blackwell Scientific Publications, Oxford, pp. 1–228.

Lewis, E.B. (1978): A gene complex controlling segmentation in *Drosophila*. Nature **276**: 565–570.

Lima-de-Faria, A. (1994): Biological periodicity with reference to higher mammals and humans. In: "Principles of Medical Biology, Volume 1B, Evolutionary Biology", JAI Press Inc., pp . 253–319.

Lima-de-Faria, A. (1995): Biological Periodicity. Its Molecular Mechanism and Evolutionary Implications. JAI Press, Greenwich, CT, USA.

Lima-de-Faria, A. (1997): The atomic basis of biological symmetry and periodicity. BioSystems, **43**: 115–135.

Lima-de-Faria, A. (1998a): The role of homeotic genes and of molecular mimicry in the determination of plant and animal symmetries. In: Symmetry in Plants. (Eds. D. Barabe and R.V. Jean), World Scientific Publishing, Singapore, pp. XXVII–XXXIII.

Lima-de-Faria, A. (1998b): The atomic origin of structural periodicity. In: Symmetry in Plants. (Eds. D. Barabe and R.V. Jean), World Scientific Publishing, Singapore, pp. 655–679.

Macdonald, D. (1984): The Encyclopaedia of mammals, Vol. 1 and 2. (Ed. D. Macdonald). George Allen and Unwin, London, Sydney, pp. 1–895.

McFarland, D. (1981): The Oxford Companion to Animal Behaviour (Ed. D. McFarland). Oxford University Press, Oxford, New York, pp. 1–657.

McGinnis, W. and Kuziora, M. (1994): The molecular architects of body design. Scientific American **270** (2): 36–42.

McMahon, T.A. and Bonner, J.T. (1983): On Size and Life. Scientific American Library, Scientific American Books, Inc., New York

Ny, T., Elgh, F. and Lund, B. (1984): The structure of the human tissue-type plasminogen activator gene: Correlation of intron and exon structures to functional and structural domains. Proc. Nat. Acad. Sci. USA **81**: 5355–5359.

Peifer, M., Karch, F., Bender, W. (1987): The bithorax complex: control of segmental identity. Genes Dev. **1**: 891–898.

Perrins, C. (1976): Bird Life. An Introduction to the World of Birds. Elsevier-Phaidon.

Pringle, J.W.S. (1975): Insect Flight. Oxford Biology Readers: **52** (Ed. J.J. Head), Oxford University Press , London, pp. 1–16.

Ryden, L.G. and Hunt, L.T. (1993): Evolution of protein complexity: The blue copper-containing oxidases and related proteins. J. Mol. Evol. **36**: 41–66.

Sanderson, R.T. (1967): Inorganic Chemistry. Reinhold Publishing Co., New York, U.S.A.

Savage, R.J.G. and Long, M.R. (1986): Mammal Evolution . An Illustrated Guide. British Museum (Natural History), London.

Smith, C.W.J., Patton, J.G. and Nadal-Ginard, B. (1989): Alternative splicing in the control of gene expression. Ann. Rev. Genet. **23**: 527–577.

Südhof, T.C., Goldstein, J.L., Brown, M.S. and Russel, D.W. (1985a): The LDL receptor gene: A mosaic of exons shared with different proteins. Science **228**: 815–822.

Südhof, T.C., Russel, D.W., Goldstein, J.L. and Brown, M.S. (1985b): Casette of eight exons shared by genes for LDL receptor and EGF precursor. Science **228**: 893–895.

Van der Ploeg, L.H.T. (1990): Antigenic variation in African trypanosomes: genetic recombination and transcriptional control of VSG genes. In: "Gene Rearrangement" (Eds. B.D. Hames and D.M. Glover). IRL Press, Oxford University Press, Oxford, New York, pp. 51–97.

Walker, C. (1974): Introduction: the world of birds . In: "The World Atlas of Birds" . Mitchell Beazley Publishers, London, pp. 10–33.

Whitten, D.G.A. and Brooks, J.R.V. (1988): The Penguin Dictionary of Geology. Penguin Books, London, pp. 1–493.

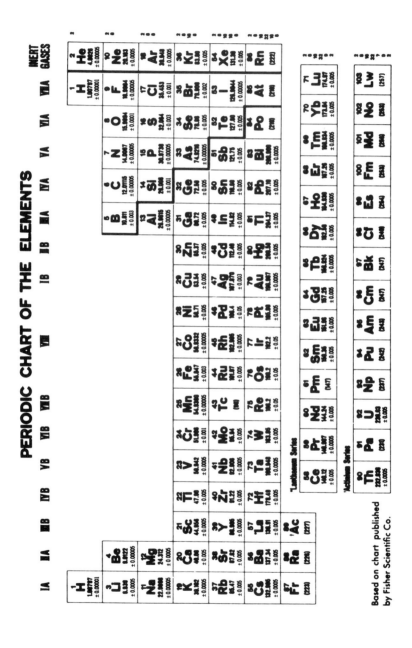

Fig. 1. Periodic table of the elements. The periodicity found among the slightly over 100 atoms known, is usually displayed as a periodic table of elements. The periodicity refers to the fact that atoms in the same group have similar chemical properties. (From Sanderson, 1967).

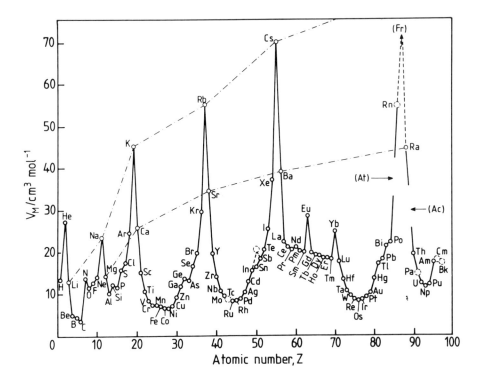

Fig. 2. Another way in which the periodicity of the elements has been displayed. Graphic representation of variation in volume per mol, as the atomic number (Z) increases. The atomic number of an element denotes the number of protons contained in the nucleus of an atom of the element. The periodicity is revealed by the peaks that occur at Na (sodium), K (potassium), Rb (rubidium), Cs (cesium), and Fr (francium). (From Greenwood and Earnshaw, 1989).

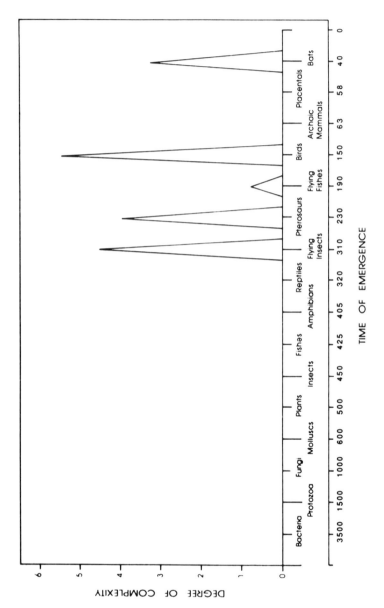

Fig. 3. Periodicity of flight. Flight is a phenomenon that arises from nowhere. There are 30 phyla of invertebrates but in only one of them did the ability to fly develop. Moreover, it attained in insects such a capacity that it was seldom exceeded later in evolution. It appears again suddenly in the reptiles called pterosaurs (pteron = wing), in flying fishes, birds, and bats which are mammals). None of these groups are directly related to the preceding group that was able to fly but have arisen from nonflying relatives. Flying fishes are teleosts; they developed after their emergence which was about 190 million years ago. The approximate time of emergence given in the figure is in millions of years. The degree of complexity is a crude estimate of the number of specialized tissues and organs involved in flight. The time of emergence is not drawn to scale to avoid making the graphs too large. (From Lima-de-Faria, 1995).

Fig. 4. Periodicity of flight: structural similarity. (1) Insect *Chrysops discalis*. (2) Extinct flying reptile *Ramphorhyngue*. (3) The bird, giant kingfisher, *Megaceryle maxima*. (4) Flying fish, *Dactylopterus orientalis* (gurnard). (5) A bat (flying mammal). Source: (1) Brumpt, 1936; (2) Boule and Piveteau, 1935; (3) Beazley, 1974; (4) Freeman, 1972; (5) Perrins, 1976.

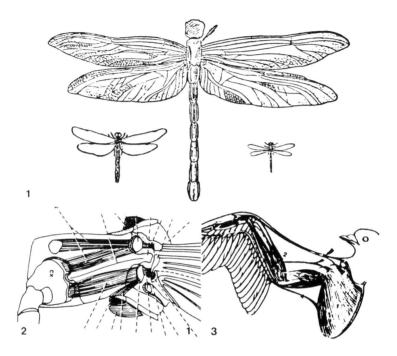

Fig. 5. Flight in insects and birds. – (1) The fossil of the giant dragonfly *Meganeura* that inhabited the forests of the Carboniferous period (350 to 270 million years ago). The wings from tip to tip measure 73 cm. For purposes of comparison two of the largest living dragonflies are depicted here: one from South America (left) and another from Europe (right). (2) The wing muscles of an insect. Four different types of muscles are involved in the wing's movement: basalar, dorsal longitudinal, subalar and tergosternal. Chitinous plates and connections lying between the muscles and the wing base are additional structures that coordinate flight movements. (3) The wing muscles of a bird. The three main types of muscles that move the wing are: pectoralis major, triceps and biceps. Several tendons are auxiliary structures that hold feathers in position. Source (1) Francé; 1943; (2) Borror et al., 1976; (3) Perrins, 1976.

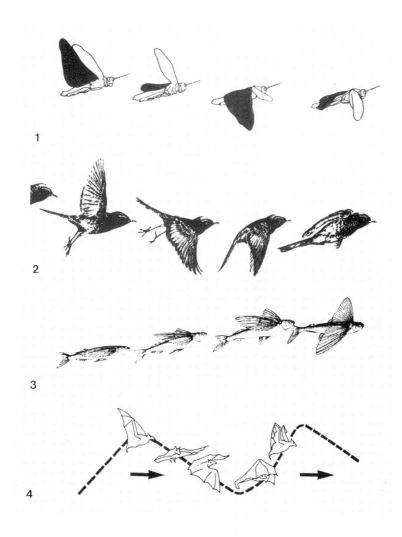

Fig. 6. Periodicity of flight – Functional similarity. (1) The course of wing movements in the locust (insect). (2) High-speed cine photography of bird flight. Take off is usually achieved by jumping into the air. (3) The flying fish *Exocoetus callopterus* which uses the propulsion of its tail for forward speed and then opens its pectoral and pelvic fins for flight. (4) Sequential positions of the bat wings during flight. Source: (1) Pringle, 1975; (2) Perrins, 1976; (3) Beazley, 1980; (4) Savage and Long, 1986.

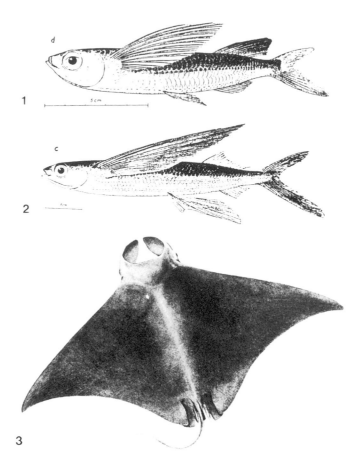

Fig. 7. Flight in fishes (1) The Atlantic two-winged flying fish *Parexocoetus mento atlanticus*. (2) The four-winged flying fish *Cypsilurus heterurus* also from the Atlantic Ocean. (3) A ray moves through the water using the wing-beat motions of birds and for this reason is considered to fly but in a different medium than air, which in this case is water. Source: (1) and (2) Hanström and Johnels, 1962; (3) McMahon and Bonner, 1983.

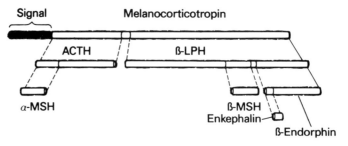

Fig. 8. One polypeptide gives rise to six hormones that activate different organs. The melanocorticotropic hormone is carved into adrenocorticotropin (ACTH) and beta-LPH (lipotropin). A second cleavage occurs releasing alpha- and beta-MSH (melanocyte stimulating hormones) and beta endorphin . The third cleavage produces enkephalin. (From de Duve, 1984).

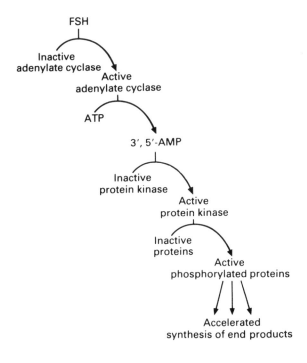

FSH

Inactive
adenylate cyclase

Active
adenylate cyclase

ATP

3′, 5′-AMP

Inactive
protein kinase

Active
protein kinase

Inactive
proteins

Active
phosphorylated proteins

Accelerated
synthesis of end products

Fig. 9. Cascade of chemical events involved in the mechanism of action of the follicle stimulating hormone (FSH). (From Dorrington, 1979).

Fig. 10. A normal fly of *Drosophila melanogaster* with its two wings. Behind the wings are the two halteres. These are clubshaped rudimentary structures which are located on the metathorax. They beat at the same frequency as the forewings and detect course changes in flight (From Alberts et al. 1994, Courtesy of E.B. Lewis).

Fig. 11. The *Drosophila melanogaster* mutant which is a four-winged fly with two forewings and two hind wings. This has been produced by combining the *bithorax* and *postbithorax* mutations by means of a rerrangement of bases in its DNA. Mitotic recombination in the chromosomes of the fly led to a deficiency in the *ultrabithorax* gene that resulted in the formation of an extra pair of hind wings (From Lawrence 1992). The extra wings have the following features: (1) They have the same structure and shape as normal wings. (2) They are located on the dorsal side of the body like normal wings. (3) They have a similar pattern of venation as the forewings. (4) The novel hind wings are slightly smaller than the forewings; this is a general feature in insects that normally have four wings, such as the butterflies. (5) The hind wings have articulations and muscles, which are obligatory components of flight, as well as access to blood flow. In this mutant what seems to partly impair flight is that some of the muscles have not become fully developed. However, research ist at present being carried out to obtain mutants with full muscular capacity. On the basis of random mutations one would have expected the appearance of a deformed structure with an accidental shape and body location, and lacking most accessory critical structures such as venations, articulations and muscles. This is not what happened. The most significant aspect from the evolutionary point of view is that perfect wings, with their normal structural and physiological components, were formed directly and in a single step. Moreover, no selection intervened in the creation of this most harmonius pattern, that emerged, structurally and functionally apt, from the very start.

The evolution of avian ontogenies –
Determination of molecular evolution by integrated complex functional systems and ecological conditions

HANS-RAINER DUNCKER

Introduction

The exploration of evolutionary processes is mostly based on investigations of the structural, developmental or functional changes of one organ system in related species, genera, families or orders either in fossils or recent organisms. These findings are arranged in sequences of the structural or functional features, the changes of which are caused by the randomly occurring mutations or other genetic changes and selected by the adaptational value for the fitness to the population, in which this structural or functional change took place. By separation of single structural or functional components from their integration into the functional systems of the organism, this interpretation ignores not only their complex functional interactions, but also neglects their close developmental interdependencies. The presented examples will explain in which way interrelationships between a larger number of different organ systems govern the selection of the randomly occurring genetic changes in their development as well as in their functions, and thus determine an effective molecular evolution, which is the basic mechanism of functional and phenomenological changes in the species during the course of evolution.

The presented examples deal with the evolution of avian ontogenies. Similar to their reptilian ancestors, all taxonomically primitive basic birds possess precocial hatchlings, probably resembling the original ontogenetic development of their avian ancestors. In the different orders of recent birds varying degrees of the evolutionary development of their ontogenetic mode towards altriciality are found. This trend toward change in the ontogenetic mode culminated in highly differentiated song birds, parakeets and parrots, all of which develop extreme altricial nestling forms. These phenomena have been lucidly demonstrated by Portmann (1936, 1938, 1939, 1954, 1959), who on the basis of the methods and results available to him at the time explained this phylogenetic problem of change from precocial toward altricial forms, giving a special interpretation of the underlying processes. His contributions described the basic evolutionary changes correctly, whereas it is difficult to follow all his explanations of these processes. Recently, Starck (1989, 1993, 1997), Starck and Ricklefs (1997a, b), Ricklefs and Starck (1997) and Ricklefs et al. (1997) thoroughly reinvestigated these problems using new methodologies by analysing the development of several organ systems in size-different precocial and altricial species. Together with the investigation of the development of further organ systems (Duncker 1978, Schepelmann 1990, Brackmann 1991, Gräf 1991, Perlich 1991, Rohrbach

1995) the results are given, upon which the presented examples of this paper are based.

Investigated Material

In contrast to the explanations given by Portmann, we found that the overall embryological development in all birds is rather rigid; the sequence and the time course of the single developmental steps is very similar in all groups of birds. The investigation of the embryological development of six species of birds including very specialized precocial species as well as highly differentiated altricial forms, demonstrated the same number and sequence of developmental stages in all species on the basis of the description and documentation of their externally recognizable characters of the developmental progress (Starck 1989). The embryological development of these species can be subdivided into 42 stages, identical for the six species. Thus, these developmental stages of the three precocial species: Moscovy duck (*Cairina moschata* L. 1758), European quail (*Coturnix coturnix* L. 1758) and barred button-quail (*Turnix succitator* Gmelin 1789) and the three altricial species: domestic pigeon (*Columba livia* Gmelin 1789 f. dom.), budgerigar (*Mellopsittacus undulatus* Shaw 1805) and Java sparrow (*Lonchura orizyvora* L. 1758) were represented in one table of normal stages (Starck 1989). The embryos of all these species hatch at stage 42, which with respect to the developmental stage of most organ systems is closely comparable, at least for the development of the skin, the skeletal-muscular system, the intestinal, respiratory and cardiovascular systems. Comparing the typical precocial hatchlings, which can immediately follow and imitate their parents, and the highly evolved naked altricial nestlings of songbirds and parrots, the visually apparent differences are seen in the development of the feathers and the functional abilities of their sensory, locomotory, behavioural and thermoregulatory systems.

For a better understanding of the structural and functional differences between these different developmental modes in precocial and altricial hatchlings, as well as to envisage the biological meaning and the evolutionary implications of these phylogenetic changes in the ontogenetic modes, the exact time tables of the structural and functional development of the skin and feather system, of the skeletal system (Starck 1989, 1993, 1997), the respiratory (Duncker 1978) and intestinal systems (Perlich 1991), or on the blood and the blood cell-producing bone marrow (Schepelmann 1990, Brackmann 1991, Gräf 1991), and the brain (Starck 1989, Rohrbach 1995) have been investigated, including the chorio-allantoic membrane (Arnoldi 1999), the embryonic gas exchange organ. The embryological development of these organ systems as well as their post-hatching growth towards adult size have been investigated. These investigations have been performed to varying degrees in the different organ systems macroscopically, histologically by light microscopy, by transmission and scanning electron-microscopy as well as by quantitative morphometric methods. From the different methods employed those which were able to answer the functionally significant questions were applied for the investigation of the single organ systems. Special emphasis will be placed on the relationships and interdependencies of the developmental time tables of the different organ systems and on their functional and evolutionary significance.

Outline of the results

The development of the outer appearance of the embryos of all investigated species with their different ontogenetic modes is very similar in regard to their recognizable anatomical features, which characterize the single embryonic stages. Also the time sequence of the developmental steps of the different organ systems with regard to these embryonic stages is almost identical in the different forms. Generally, the embryonic development of all investigated species can be subdivided into four developmental phases. The first phase is made up by the development of the basic organization of the body, which extends up to stage 22. It is followed by the second phase of the development of the organ anlagen up to stage 33. The following phase of differentiation of the organs continues up to stage 38. The fourth embryonic phase is occupied by late embryonic growth (fig. 1) and adaptation for hatching at stage 42 (Starck 1989, 1993, Ricklefs and Starck 1997). These embryonic phases are followed by the posthatching growth phase, in which the adult body size is reached at very variable lengths of time, 80 to 120 days in precocial species up to 180 days or more, whereas in altricial birds adult size is reached in most species within 25 to 30 days (Starck 1989, Ricklefs and Starck 1997, Duncker 1998).

In the development of the skin the first anlagen of the feathers are formed subcutaneously at stage 31 in all investigated species. This development of the first plumage is followed in precocial birds by the outgrowth of the feather anlagen beyond the surface of the skin at stage 35 so that at hatching these species possess a fully developed down-plumage (fig. 2), and they also show a rather highly developed subcutaneous system of feather and apterial muscles and tendons. After hatching, the plumage continues its development, giving origin to the subsequent generations of contour feathers. In the highly evolved altricial species development of the feather anlagen is retarded compared to typical precocial embryos and the feathers do not grow beyond the surface of the skin. However, even with this retardation in the development of the feathers the subcutaneous system of feather and apterial muscles and tendons has reached the same developmental level as that in precocial embryos (Starck 1989, 1993, Ricklefs and Starck 1997). Due to this change in the time table of feather development, the typical altricial nestlings are naked at hatching. Regularly, their first feather generation, the down plumage, grows beyond the skin surface after the eighth or tenth nestling day (fig. 2), then develops rapidly and soon is followed by the next generations of feathers. The fact that this developmental time table is caused by a retardation in the feather development and not by a lack of available time, as suggested by Portmann (1938, 1959) to explain this phenomenon, is demonstrated, for example, by the hatchlings of some typical song birds such as the thrushes, which are generally naked at hatching but possess on their heads and backs a few signal feathers, which are fully developed natal downs.

The development of the skeletal elements demonstrates very similar features to those of the skin-feather system. In all investigated species the formation of the cartilaginous elements starts at stage 27 and continues in a very comparable way in precocial and altricial species with very similar time sequences of the commence-ment and growth of the different cartilages (fig. 3). The beginning of ossification takes place at stage 33, including the peri- and enchondral ossification of the carti-laginous elements of the trunk as well as the primordial cranium and the dermal bones. This ossification continues rapidly in precocial embryos so that at the time of hatching the greatest parts of their skeletal elements are ossified according to the biomechanical needs of their sustained locomotory activities directly after hatching (fig. 3). Consequently, their skeletal elements contain only small distal regions con-sisting of cartilage, which are responsible for the relative slow longitudinal growth of the skeleton and the limited body growth of the hatchlings. In contrast to the precocial species, in altricial birds the ossification continues more slowly after its onset. By this retardation the extent of ossification of the cartilaginous elements at hatching in altricial nestlings is remarkably less compared to precocial hatchings, especially in the vertebral column and in the bones of the legs and wings, which possess large proximal and distal cartilaginous parts (Starck 1989, 1996a, 1997). Thus, in altricial nestlings the skeleton, which does not have to serve greater loco-motory activities, is highly adapted for the acclerated longitudinal growth and the general rapid body growth of the nestlings by virtue of the retarded ossification and thereby by the large cartilaginous parts, consequently reaching adult body size with-in 20 to 30 days (fig. 3).

    The development of the respiratory and intestinal systems demonstrate in pre-cocial and in altrial embryos the same general structural and temporal patterns in the second and third embryonic phases of the development of organs and their differen-tiations (Duncker 1978, Perlich 1991). Only at the end of the third embryonic phase and during the fourth phase of late embryonic growth both systems develop quanti-tatively very differently. In precocial embryos the lungs demonstrate in the last phas-es a strong thickening of the parabronchial walls containing the three-dimensional network of blood and air capillaries. Through air breathing by the embryo in the egg the lung and its air capillaries are aerated in the last one to two days of the incubation period. The precocial hatchlings leave the egg with a rather large gas exchange sur-face, which supports their high locomotory and maintenance metabolism. In con-trast, in altricial embryos the growth of the parabronchial walls in the last embryonic phases is very limited, such that at hatching they possess a thin layer of air and blood capillaries in their parabronchial walls, sufficient for the support of their relatively low maintenance metabolism (Duncker 1978). In contrast to the respiratory system, the intestinal tract develops in the last embryonic phases in small altricial species an absorptive surface in the small intestine that is four times larger per gram hatchling body weight than in precocial species (Perlich 1991) (fig. 4). These differences dem-onstrate a prerequisite for the extreme increase in growth metabolism of altricial nestlings, which enables them to perform their extremely rapid body growth, com-pared with precocial hatchlings, which have to support mainly their maintenance metabolism and only a low growth metabolism with their limited small intestine absorptive surface. But considering the data obtained from the weight of the intes-tine in relation to body weight of the hatchlings in different precocial and altricial species, especially including large altricial species, the results are inconclusive in respect to these precocial-altricial differences in gut size (Starck 1997). Both the maintenance metabolism in precocial hatchlings, depending on their small body size, as well as the high growth metabolism in altricial nestlings, surpass the metabolism of their adult forms by a factor of two or more. Correspondingly, the blood volume and especially the mass of the erythropoetic bone marrow increases rapidly after hatching in the three investigated species quail, pigeon und budgerigar during the growth period up to 2.5 times of the mass in adult birds relative to their body mass, being most rapid in budgerigars (Schepelmann 1990, Brackmann 1991, Gräf 1991), thus serving the necessary blood transport capacity together with an increased general blood volume (Wagner 1985).

The most striking differences in the ontogenetic development of precocial and altricial species are seen in the development of the brain (fig. 5). Precocial hatchlings emerge from the egg with a fully developed brain stem and also with a highly developed forebrain (Starck 1989, 1997, Rohrbach 1995). They possess not only the ability to regulate all their vegetative functions and even thermoregulation soon after hatching, but they also have a forebrain, which in all its different regions is cytologically differentiated and completely functionally competent. Directly after hatching they are imprinted by the voice and appearance of their mother; they perform all necessary locomotory and behavioural activities from gathering their food up to hiding from enemies. In contrast, at hatching in the highly evolved altricial nestlings only the brain stem is functionally differentiated, regulating their vegetative functions including the few behavioural reflexes that are necessary to be fed by their parents. At hatch, their eyes are closed and they are incapable of vision. The forebrain of these altricial nestlings is still fully proliferating, which comes to an end ten days after hatching. Only then the cytological differentiation of the different striatal areas of the forebrain can start so that around 20 days after hatching their forebrain reaches the functional capacity that the forebrain of precocial hatchlings performs already at hatching (Starck 1989, 1997). Thus, 20 day after hatching the altricial nestlings begin to imprint their species-specific song, food and brood care behaviour. Reaching adult body size within the next few days, these altricial nestlings become fledglings which leave the nest.

Brain growth after hatching in precocial species reaches maximally a weight 3.6 times that at the time of hatching (fig. 5), mainly due to myelinization. In pigeons, which show a remarkable proliferation in the forebrain during the first ten days, the overall brain growth from hatching to adult size is 6.6 times. In the altricial budgerigar the overall brain growth from hatching to adult is 15 times, in the Java sparrow 16.3 times (Starck 1989, 1993). Both ontogenetic modes of development in precocial and altricial species start at hatching with a very similar brain weight/body weight relationship. Only the pigeon demonstrates a much lower brain/body weight relationship. As a result of the strongly increased forebrain proliferation and growth in altricial species (budgerigar and Java sparrow) a number of their striatal forebrain complexes possesses even a growth factor from hatching to adult size amounting to 25 to 30 times (Starck 1989, 1993). These data demonstrate the developmentally high level of cerebralization in the highly evolved altricial orders of the songbirds and parrots.

Functional consequences of ontogenetic developmental differences

The differences in the development of the investigated organ systems, which occur within the rather rigid and strict general ontogenetic developmental scheme of birds have been outlined here using the examples of three precocial species, an intermediate altricial form and two altricial species. The functional consequences of these developmental differences will be explained for the highly different precocial and altricial modes (fig. 6). Nevertheless, looking at the ontogenetic development in a larger number of avian species and orders, they demonstrate a continuous sequence of ontogenetic modes, which have been subdivided into eight ontogenetic types, four precocial types, one intermediate type (Platzhocker) and three altricial types (Starck 1989, 1993, Starck and Ricklefs 1997). Looking at the embryonic and post-hatching development of single organ systems and functional abilities of these birds belonging to these different ontogenetic types, we will find continuous sequences of differentiation at certain developmental stages and at hatching; thus, for example, from a fully developed down plumage in precocial hatchlings to totally naked song-bird hatchlings, which in altricial species results from a retardation of feather development by an alteration in the developmental time table. In the same way the development of the other organ and functional systems are modified by alterations in their developmental time tables, as illustrated by the outlined examples: The differences in the ossification of the skeleton at hatching, the differences in the amount of exchange surface in the lung parabronchi or the differences in the small intestinal absorptive surface are due to such alterations in the proliferation and growth time tables of these organ systems. Most pronounced are those alterations in the proliferation and growth time table of the brain between the investigated precocial and altricial species.

Depending on or connected with these developmental time table alterations a large number of functional systems are also modified in the time course of their development. Precocial hatchlings imprint to the voice and visual appearance of their mother directly after hatching. Altricial nestlings perform the learning process of the species-specific song, food and nesting material around 20 days after hatching. Precocial hatchlings gain their thermoregulatory capacity soon after hatching, altricial forms 10 or more days after appearance of the plumage. Precocial hatchlings are locomotory competent at hatching, altricial forms gain this competence not until the process of fledging. Precocial hatchlings support their maintenance metabolism in the first four days or more post hatching via the yolk sac content, especially in larger species, whereas smaller species can support their metabolism by the relative smaller yolk reserve only for a shorter time, being forced to take up food sooner. Altricial hatchlings do not possess yolk reserve and have to be fed by the parents directly after hatching with large amounts of food to maintain their high metabolism. Their metabolism is only to a very limited extent necessary for maintenance, the majority of it being invested into growth. In contrast, precocial nestlings expend most of their metabolism for maintenance, locomotion and thermoregulation, whereas only a limited amount is available for growth. These relations determine the growth curves and growth rates of precocial hatchlings, and explain the relatively long time of 80 to 120 days or more needed for reaching adult body size (fig. 6). In contrast, altricial nestlings, at least in small and medium-sized species, attain adult body size in 22 to 30 days via their extremely accelerated growth rates (Starck 1989, 1993, Starck and Ricklefs 1997). This is explained by the fact that most of the metabolism subsidizes growth and only a limited amount is needed for maintenance, especially in the early nestling phase, when they lack the ability of thermoregulation, depending totally on warming through parental brood care. The structural basis for this accelerated growth is given by the enlarged intestinal absorptive surface and transport capacity and by the relatively large cartilaginous parts of the skeletal elements for longitudinal growth.

All these alterations in developmental time tables of growth processes of single organ systems or of single functional systems and abilities can be arranged for each system in continuous sequences of modifications, correlating all discovered alterations in the different bird orders with their different developmental modes. However, this view of the phenomena explains only the way in which evolution could have altered these developmental processes of single systems, isolated from the living species. Contemplating of a single species with its ontogenetic development, we have to concentrate on quite another aspect, to obtain the biologically important correlations that determine the specific ontogenetic mode. In a single species the developmental time tables of its different organs and functional systems are arranged in a very specific and highly differentiated way and very precisely adapted to one another. Fully developed plumage at hatching, full locomotor activity and advanced ossification of the skeleton, full sensory and behavioural competence and functionally fully differentiated sense organs and a brain with a functionally competent forebrain, high maintenance metabolism and limited growth rate with a relatively long time for acquiring adult body size are all closely interlinked with one another, each depending on the others in their specific degree of development and activity. The same is true for the differentiation and specific developmental time courses of the various organs and functional systems in altricial species. The strongly reduced locomotor activity of their hatchlings is correlated to the very limited function of the sense organs and the regulation by the brainstem of the required autonomic functions and the few behavioural reactions, allowing intensive proliferation and later cytological differentiation of the forebrain and of the eye. The extensive small intestinal absorptive surface, the intensive feeding behaviour by the parents, the large blood volume and the long cartilaginous portions of the skeletal elements all together are responsible for the extremely accelerated growth, and are supported by the lack of a large maintenance metabolism and the total lack of thermoregulation so that most of the energy from food uptake is available for growth. Being naked in the first nestling phase is a biological adaptation to the intensive warming by the parents after their return from food gathering. The first plumage develops when the parents are forced by the increased food requirement of the growing nestlings to spend more time for acquiring food. At this time thermoregulation also develops in the nestlings. Thus, by these special arrangements in the specific design of the developmental time tables of the different organs and functional systems the accelerated growth leading to adult body size is attained concomitant with a high degree of cerebraliza-

tion and an extemely enlarged forebrain for highly evolved behavioural and locomotor performance.

In this way the different developmental time tables for each organ are arranged
in a very specific time pattern in each single species. In birds there are not only these
two highly different developmental time patterns for precocial and altricial species,
but each species or group of species possesses its very own specific arrangements of
these developmental time tables making up their species-specific developmental
time pattern. These time patterns of the different genera and orders of birds are not
only determined by the internal conditions of the various organ systems of these
homoiothermic vertebrates, but also by the overall ecological conditions under which
the species are living and reproducing successfully. Thus, the environmental, climatic and seasonal conditions, the availability of food, the size of the clutch, but
also hatching and adult body size of the species and its food specialization are factors to which the specific developmental time pattern of each species is adapted. As
well, the sequence of egg laying and commencement of incubation and the resulting
sequence of hatching are functionally very important factors. This range of differentiation can reach from an absolute synchronization, e.g. in some ducks, incubating
their eggs on rocks or in trees, up to hatching occurring over several days, e.g. in
birds of prey, which determines the very different nestling body sizes and highly
different survival probabilities, and thus may represent a flexible adaptation of certain species to short term changes in food availability (Ricklefs et al. 1997). In these
ways the reproductive strategy of the single species is very specifically adapted, on
the one hand, to their very special overall environmental conditions, and, on the
other hand, determined by the very specific internal conditions of the developmental processes as a result of the phylogenetic developmental of the different avian
orders.

Developmental Preconditions in the Evolution of Homoiothermy determined by the Integration of Functional Systems

For a better understanding of the arrangement of the various time tables in the development of the single organs and functional systems into the species-specific developmental time patterns, and their changes during evolution of the different ontogenetic modes, they should be viewed within the framework of the complex integration of the different functional systems that have been evolved during phylogenetic
development of homoiothermy (Duncker 1991a, b, 1992). These developments were
based on the improvement of a sustainable aerobic capacity of the locomotor apparatus, which was supported by a parallel improvement in the cardiovascular and
respiratory systems (Duncker 1989). The performance of these systems as well as
the cellular metabolism and transport rates throughout the body are characterized by
the fact that the rate of transport or turnover of substances can be increased maximally tenfold: If it is necessary to attain maximal performance of these systems for
a specific biological purpose, such as sustained running or flying, this is only possible by increasing the basal functional rates of these systems by increasing the basal

metabolic rate (Duncker 1991a, b, 1992). The high basal metabolism of homoio-
therms presupposes not only high transport rates over all cell membranes and high
activities of the membrane protein transport system, but it also determines high
growth rates. The latter are the basis for embryonic and posthatching development
in birds. An outline of the results and the functional consequences of these developments demonstrate in which way the metabolic and transport rates of the different
organ systems of homoiothermic birds are arranged within general ontogenetic development. By virtue of an extensive absorptive surface in the intestine and subsequent high growth rates as well as special alterations of the time table of forebrain
and body weight development, altricial hatchlings are able to achieve a high degree
of cerebralization and to attain full adult body size and complete functional performance within a relatively short time period (Duncker 1998).

The history of the phylogenetic development of land vertebrates, especially that
of the birds, determines a large number of structures and functional abilities of the
avian body, but particularly a large number of correlations of these structures and
functions, as already stated by Cuvier in his correlation theory at the end of 18th
century. The structures that make up an avian body, and the general ontogenetic
sequence of their development in the described embryonic phases are very rigidly
fixed. On this basis, evolution of homoiothermy during the phylogenetic development of birds took place and was characterized by a multiple integration of numerous structural and functional systems, which facilitated a sustainable aerobic capacity for the locomotor apparatus depending on extremely high exercise rates, which
are based on strongly increased basal metabolic rates. Moreover, these conditions
led to a shortened embryonic development and at least in altricial birds a rapid post-
hatching development. These multiple structural and functional correlations of the
various organ systems and their functional interdependencies are the limiting factors for evolutionary changes, because all these historical and functional structures
and their mandatory correlations must be maintained to guarantee the functionality
of the avian body (Starck and Ricklefs 1997). Thus, the potential further evolutionary development of birds can only proceed with very few alterations in structure,
function and ontogenetic development. These general characteristics of phylogenetic development have already been described in different classes of organisms by
biologists of the last century, and they have termed the phenomenologically described evolutionary trends as "epigenesis" or "orthogenesis". Today, we can explain these general trends by our knowledge of the complex structural and developmental integrations of the organ systems and their functional interdependencies. As
described in the presented examples, an investigation of the evolution of the different modes of avian ontogeny demonstrated that these evolutionary developments
with their phenomenologically impressive changes from the precocial to the highly
altricial developmental mode occurs almost exclusively by altering the time tables
of the development of organs and functional systems. This has been termed heterochrony for the evolution of single organ systems out of different classes of organisms by McKinney and McNamara (1991) and others. However, the remarkable
result of the investigations presented here is the fact that the alterations in the time
tables of the different organ developments are not randomly arranged to make up

new developmental phenotypes, but rather all the different developmental time tables are functionally very precisely arranged to one another in one species, in accord with the species or family-specific time pattern. This is not only true within the given examples of precocial and highly altricial developmental modes, but also for all intermediate ontogenetic modes and even for the less obvious differences from family to family or species to species.

The historically developed, mandatory structural and functional correlations of the different organ systems characterize the bauplan of birds. The multiple integrations of the major functional systems constitute their homoiothermy, and the family- or species-specific developmental time patterns determine the ontogenetic mode of the specific group of birds. Taking into account these different levels of constraints (Starck and Ricklefs 1997, Ricklefs et al. 1997), the question arises in which manner these developmental time patterns evolved and how changes in the function and structure of single species were established. The answer to the first question is given by the results of evolution itself. The phylogenetic development from precocial to highly altricial species is seen in the existing sequences of different intermediate ontogenetic modes with a large number of smaller or greater variations in their ontogenetic development. However, these phylogenetic developments are not based on single heterochronic changes, or on the alteration of single organ developmental time table, but rather only by coordinated alterations in the developmental time patterns of the related species. For example, an alteration in the time when the first plumage develops in altricial nestlings has to be correlated not only with the physiological development of thermoregulation, but also with general body growth rate and food requirements. Changing the growth rate depends on parameters of organ and tissue composition and reactivity, exemplified by the correlation of intestinal absorption surface, blood transport capacities and skeletal longitudinal growth velocity, which depends on the available length of the proliferating cartilaginous portion of growing bone. These numerous alterations in the internal development of the nestling, only a few of which from several existing correlations have been mentioned, have to be correlated with the behaviour of the parents. The latter is determined by the special ambient seasonal conditions such as environmental temperature and food availability, thereby influencing for the parents the time balance between search for food and warming of the nestlings. Further important parameters are body size of the species, nutritional specialization, size of the eggs and nestlings and clutch size. When discussing evolutionary development, these multiple functional and structural correlations and interdependencies have to be taken into account.

All these functional and structural systems and their multiple correlations possess a certain amount of functional flexibility. Through this flexibility the populations are able to adapt to a certain degree of variability of the environmental conditions. This adaptation to ambient ecological changes can be performed to a certain extent by structural and functional modifications in the design of the organ systems, which are all included in the genetically based bauplan of the species and the time pattern of its developmental origin. The extent of these functional flexibilities and functional/structural modifications is known only from a few single observations.

Very few investigations have been done (Starck 1996b, Schew and Ricklefs 1997) or are currently underway. Including this knowledge of the flexibilities and potential for modification of the functional systems, a discussion of evolutionary changes in the mode of ontogenetic development must take into account the following basic conditions: The multiply integrated functional developmental systems with all their mandatory time-pattern correlations possess only very restricted possibilities for further evolution. Only limited alteration of these developing structures and functions, which do not disturb the complex functional interrelationships, is compatible with survival of the species. Only very few of these alterations can under certain environmental circumstances be more effective functionally and thereby increase the fitness of the species. As investigations have demonstrated, these alterations have been mainly or exclusively realized in the form of heterochronic changes of organ developmental time tables and thus by altering the general ontogenetic time pattern. As discussed above, such an alteration requires a coordinated change in a larger number of these complexly integrated functional organ developments. The only interpretation compliant with the consequences of these interrelationships is that the necessary complex integrations of the functional systems determine further evolution by selecting only a very few molecular genetic alterations out of a vast number of regularly occurring genetic changes. Multiple mutations and other genetic alterations, which are more or less constantly occurring (Alberts et al. 1994, Hennig 1995), represent the basic mechanism, out of which the very few changes that are compatible with the complexly integrated functional systems and thus with the survival and further evolution of the species are maintained in the population genome of the following generations.

Proposed Organismic Model of Evolution

The cited examples of the evolution of avian ontogeny from precocial to altricial hatchlings including the posthatching growth illustrate the principles of phylogenetic developments by alterations in their development time patterns. These examples were presented by describing the extreme endpoints of these phylogenetic developments. Looking at the great diversity of birds and their modes of ontogeny, we find a stepwise variation from precocial to altricial hatching forms and posthatching growth modes. Each of the discussed changes in the developmental time tables of the different organ systems occur only in minute steps, providing a more or less gliding sequence of variance from species to species: However, these stepwise changes did not occur isolated. Depending on the complex and very specific functional integration of the different organ developmental time tables in each species these changes can only occur in correlation with the different functional systems. This means that the overall time pattern of development in a species must be transformed into a new functioning time pattern with specifically altered, different developmental organ time tables, all of which guarantee a successful or even more efficient developmental time pattern than in the originating species. These conditions, which are governed by the mandatory interactions of the integrated complex functional systems, determine the orthogenetic tendency of these evolutionary changes. These changes present a new possibility for the survival of the population of this species, because the altered developmental time pattern depends on its very specific adaptation to the special way of life and to the special ecological conditions of this species for a more auspicious survival of the species. The biological meaning of these interdependencies lies in the fact that not only the complex internal functional correlations of the different organ developments and their very specific time patterns determine the state and further evolution of the population of this species, but in a similar way also the special arrangements of the environmental conditions under which this population is living, from the climatic and seasonal conditions to the spatial structure of the habitat, to the availability of food and to the pressure of preying animals. Thus, the specific functional structure of a species and its mode of development as well as its general reproductive behaviour are formed by the intensive interaction of internal functional and developmental constraints within the framework of the multiplicity of ecological factors, to which the population is exposed.

From the concept that the highly integrated, complex functional systems of the organism and their developmental time pattern determine which of the few genetic changes out of a vast myriad of regularly occurring mutations and other genetic changes will be selected and incorporated into the genome of the subsequent population, the following picture arises: Which of the few genetic changes are constructively incorporated into the population genomes depends especially on their interaction with the vast number of environmental factors in addition to the internal constraints. Alterations in these interactions with the environment, particularly alterations in single environmental factors, will change the basic conditions that determine, for the developing new generations of individuals, their selection of those very few genetic changes out of the basic stream of molecular genetic events, which produce successful alterations in the developmental time pattern and functional structures of the growing hatchlings and adults. Thus, the complex interactions between the internal functional possibilities of the individuals of a population with their environmental conditions and environmental changes are responsible for the determination of further evolution. The accidental mechanism of the molecular genetic clock serves only the broad material basis of genetic possibilities, from which only single genetic changes are incorporated into the population genome by the above-stated interactions. Most of the molecular genetic changes are eliminated by selection. Thus, the functional mechanisms for producing molecular genetic alterations by mutations or other genetic changes and the mechanisms for selection of very few of these genetic changes via the highly differentiated interactions between the complex internal functional conditions and the multiple environmental factors are two very different levels in the functional organization of organisms. In addition, the rather constant clock of molecular genetic changes and the phenotypical functional and morphological evolution, which produce the multiplicity of organisms with their diversity of hierarchically organized forms, are two very different processes, which are only very loosely connected with each another. In contrast to the constant molecular genetic clock, phenotypical evolution has a highly variable time scale.

The interactions of the complexly integrated developmental and functional systems of organisms with their environmental conditions and ecological changes de-

termine which very few of the steadily produced genetic changes are positively selected, and thereby this specific selection directs the course of phylogenetic development. Thus we have to imagine that the long-term constancy of a species, but also
the evolutionary change of such species are very precisely controlled by the interactions of the internal functional integrations with their environmental interrelationships. However, this implies a picture of a very dynamic species concept and also a
concept of a very flexible time course of evolution. The interaction of internal organismic and external environmental conditions over a long time in a constant environment is responsible for the long-term constancy of the species, whereas a change
in environmental factors can cause within a very short time period an evolutionary
change in the species and its ontogenetic development. This is demonstrated, for
example, by the adaptation of organisms to industrial environmental pollution and
especially by the breeding of farm and house animals with drastic changes in their
structural and physiological properties over only a very few generations.

Under consideration of these aspects the very different time scale of structural
and functional changes in evolution becomes more understandable. The evolution
of land vertebrates over the last 250 million years was characterized by a very specific time course: in intervals of approximately 25 to 50 million years a more or less
drastic reduction in the number of species occurred over a relatively short period of
time (Stanley 1988), after which time a new radiation occurred also over a rather
short period of time with the evolution of multiple new orders, families and species.
After the short radiation time, in which a drastically accelerated phylogenetic development took place, the newly evolved fauna demonstrated a great constancy at the
taxonomic level of its families over a period of more than 20 million years with only
long-term evolution of its species. As a result of the reduction phase, the remaining
species invaded new environments and a large number of open niches. In interaction
with these new environmental conditions and ecological factors the evolution of
development and structure of the remaining species increased dramatically within a
short period, producing new families and orders and thus radiating into the newly
conquered environments. After this rapid evolution, producing a rich diversity of
new fully adapted forms, a new stability in the interactions of the organisms with
their environments was established, which only allowed a very slow time scale of
further evolutionary changes. In this way the very different processes of the time
constant molecular genetic clock and the very variable time scale of the phylogenetic development of species, families and higher orders of organisms can be brought
together into an organismic model of evolutionary development.

Summary

The investigation of the evolution of the modes of ontogeny in birds has shown that
the development of the different organs and functional systems is very precisely
integrated one with another, not only in their general arrangements, but especially
with their single time tables, which are coordinated in species-specific developmental time patterns. This is an expression of the highly integrated general bauplan of

birds as well as of the very specific integration of the avian functional systems, which constitute their homoiothermy. These mandatory structural and functional complexes set the framework, within which the very rigid avian ontogenetic development can take place. Within this framework of functional conditions the evolution in birds and their ontogenies occur exclusively by alteration of developmental time tables of organs and functional systems, through which their highly interdependent arrangements allow only very few but strongly interconnected variation: Thus, these developmental time tables are coordinated in the single species into a species-specific time pattern. In the different species or species groups their specific ontogenetic time patterns demonstrate the very few avenues in which the evolution of avian ontogenies were able to proceed during phylogeny.

These modes of evolutionary development are determined by highly integrated structural and functional systems of ontogeny and by the adult structure of the different avian orders and species. Functional structure and special ontogenetic development are very specifically coordinated to the general and special environmental conditions. Both internal complex functional interdependencies of the single species and the multiplicity of interacting ecological factors set the framework in which evolution can occur. This means that from the more or less time-constant stream of molecular genetic changes only very few have a chance to be incorporated into the genome of the population. These include those which do not disturb the complex developmental and functional interdependencies or, in seldom cases, are able to increase the efficiency and fitness of the population. Thus, the complex internal conditions, on the one hand, and their steady interactions with all the ecological factors, on the other hand, determine, which few of the molecular genetic events are selected to govern further evolution.

These consequences constitute the fundament for a very flexible species concept: The species in its actual appearance is determined by the historical development of its hierarchically organized, highly integrated functional systems as well as by the interactions of these systems with the multiplicity of ecological conditions. Thus evolutionary changes and especially their time course are dependent on both complex systems. When the interactions of the internal conditions with the multiplicity of ecological factors are relatively constant for long periods of time, species, families and orders will also demonstrate stability over long time periods. On the other hand, short-term changes in ecological factors will cause evolutionary development of new forms, species and higher genera within rather short time periods, as shown by the radiation periods in vertebrate evolution or in the domestication of animals. The rather constant stream of molecular genetic changes and the phenotypical, morphological and functional evolution are two highly different processes, acting on very different levels of the organisms. Determined by the complex internal conditions of organisms and their interactions with the multiplicity of ecological factors, these mandatory functional interrelationships of the hierarchically organized organisms are responsible for organismic evolution, as demonstrated by the example of avian ontogenies. The results of these investigations propose a broader and flexible species concept and an organismic model of evolutionary development.

Acknowledgements: The author acknowledges with gratitude Ch. Thiele for the competent computer-aided illustrations, Dr. R. L. Snipes for linguistic revision of the text and Ms E. Hellmann for careful secretarial production of the final versions of the manuscript.

Literature

Alberts, B., Bray, D., Lewis, J., Raff, M., Roberts, K. & Watson, J.D. (1994): The Molecular Biology of the Cell. 3rd. edit. Garland Publ. New York.

Arnoldi, B. (1999): Die Entwicklung der Chorioallantoismembran von Moschusente und Wellensittich. Qualitative und quantitative-morphometrische Untersuchungen. Vet.-Med. Diss. Giessen.

Brackmann, F. (1991): Das erythropoetische Knochenmark der europäischen Wachtel: Verteilung und Volumen bei Wachstum und Pneumatisation des Skeletts. Vet.-Med. Diss. Giessen, 1–131.

Duncker, H.-R. (1978): Development of the avian respiratory and circulatory systems. in: Piiper, J. (ed.) Respiratory Function in Birds, Adult and Embryonic. Springer, Berlin, 260–273.

Duncker, H.-R. (1989): Structural and functional integration across the reptile-bird transition: Locomotor and respiratory systems. In: Wake, D.B. & Roth, G. (eds.) Complex Organismal Functions: Integration and Evolution in Vertebrates. Wiley and Sons, Chichester, 147–169.

Duncker, H.-R. (1991a): Constructional and ecological prerequisites for the evolution of homeothermy. In: Schmidt-Kittler, N. & Vogel, K. (eds.) Constructional Morphology and Evolution. Springer, Berlin, 331–357.

Duncker, H.-R. (1991b): The evolutionary biology of homoiothermic vertebrates: the analysis of complexity as a specific task of morphology, Verh. Dtsch. Zool. Ges. **84**, 39–60.

Duncker, H.-R. (1992): Die stammesgeschichtliche Entstehung von Komplexität im funktionellen Aufbau der Organismen. Mitt. hamb. Zool. Mus. Inst. **89**, Ergbd. **1**, 73–96.

Duncker, H.-R. (1998): Funktionsmorphologie und Genetik: Die Steuerung der molekularbiologischen Evolution durch die Gesetzmäßigkeiten der komplex verknüpften Funktionssysteme, dargestellt am Beispiel der Evolution der Vogelontogenesen. Theory Bioscienc. **117**: 42–77.

Gräf, F. (1991): Das erythropoetische Knochenmark des Wellensittichs: Verteilung und Volumen bei Wachstum und Pneumatisation des Skeletts. Vet.-Med. Diss. Giessen, 1–159.

Hennig, W. (1995): Genetik. Springer, Berlin, Heidelberg.

McKinney, M.L. & McNamara, K.J. (1991): Heterochrony. The Evolution of Ontogeny. Plenum Press, New York, London.

Perlich, E. (1991): Der zeitliche und quantitative Ablauf der Ontogenese des Darmtraktes von Europäischer Wachtel und Wellensittich. Biol. Diplom-Arb. Giessen, 1–124.

Portmann, A. (1936): Die Ontogenese der Vögel als Evolutionsproblem. Verh. Schweiz. Naturforsch. Ges., 224–241.

Portmann, A. (1938): Beiträge zur Kenntnis der Postembryonalentwicklung der Vögel. Revue Suisse Zool. **45** (6), 273–348.

Portmann, A. (1939): Nesthocker und Nestflüchter als Entwicklungszustände von verschiedener Wertigkeit bei Vögeln und Säugern. Revue Suisse Zool. **46** (12), 385–390.

Portmann, A. (1954): Die postembryonale Entwicklung der Vögel als Evolutionsproblem. Acta XI. Congr. Int. Orn. (Basel 1954), 138–151.

Portmann, A. (1959): Die Entwicklungsperiode vom 11. bis 14. Bruttag und die Verkürzung der Brutzeit bei Vögeln. Vierteljahresschrift Naturforsch. Ges. Zürich **104**, 200–207.

Ricklefs, R.E., Starck, J.M. & Konarzewski, M. (1997): Internal constraints on growth in birds. In: Starck, J.M. & Ricklefs, R.E. (eds.) Avian Growth and Development. Evolution within the Altricial-Precocial Spectrum. Oxford Univ. Press, New York, 266–287.

Ricklefs, R.E. & Starck, J.M. (1997): Embryonic growth and development. In: Starck, J.M. & Ricklefs, R.E. (eds.) Avian Growth and Development. Evolution within the Altricial-Precocial Spectrum. Oxford Univ. Press, New York, 266–287.

Rohrbach, St. (1995): Die postnatale Gehirnentwicklung der chinesischen Zwergwachtel (Excalfac-
    toria sinensis L.). Med. Diss. Giessen, 1–95.

Schepelmann, K. (1990): Erythropoietic bone narrow in the pigeon: Development of its distribution
    and volume during growth and pneumatization of bones. J. Morphol. **203**, 21–34.

Schew, W.A. & Ricklefs, R.E. (1997): Developmental plasticity. In: Starck, J.M. & Ricklefs, R.E.
    (eds.) Avian Growth and Development. Evolution within the Altricial-Precocial Spectrum.
    Oxford Univ. Press, New York, 288–304.

Stanley, St.M. (1988): Krisen der Evolution. Artensterben in der Erdgeschichte. Spektrum d. Wiss.
    Verlag, Heidelberg.

Starck, J.M. (1989): Zeitmuster der Ontogenesen bei nestflüchtenden und nesthockenden Vögeln.
    Cour. Forsch.-Inst. Senckenberg **114**, 1–319.

Starck, J.M. (1993): Evolution of avian ontogenics. Current Ornithology **10**, 275–366.

Starck, J.M. (1996a): Comparative morphology and cytokinetics of skeletal growth in hatchlings of
    altricial and precocial birds. Zool. Anz. **235**, 53–75.

Starck, J.M. (1996b): Phenotypic plasticity, cellular dynamics, and epithelial turnover of the intes-
    tine of Japanese quail (Coturnix coturnix japonica), J. Zool. Lond. **238**, 53–79.

Starck, J.M. (1997): Structural variants and invariants in avian embryonic and postnatal develop-
    ment. In: Starck, J.M. & Ricklefs, R.E. (eds.) Avian Growth and Development. Evolution with-
    in the Altricial-Precocial Spectrum. Oxford Univ. Press, New York, 59–88.

Starck, J.M. & Ricklefs, R.E. 1997a Patterns in development: The altricial-precocial spectrum. In:
    Starck, J.M. & Ricklefs, R.E. (eds.) Avian Growth and Development. Evolution within the
    Altricial-Precocial Spectrum. Oxford Univ. Press, New York, 3–30.

Starck, J.M. & Ricklefs, R.E. (1997b): Variation, constraint, and phylogeny: Comparative analysis
    of variation in growth. In: Starck, J.M. & Ricklefs, R.E. (eds.) Avian Growth and Development.
    Evolution within the Altricial-Precocial Spectrum. Oxford Univ. Press, New York, 247–265.

Wagner, R. (1985): Veränderung von Blutvolumen, Erythrokinetik und Eisenkinetik während der
    postnatalen Entwicklung von Haushühnern. Med. Diss. Giessen, 1–87.

Temporal sequence of embryonic and postnatal growth in precocial and altricial birds

Body size given as relative values

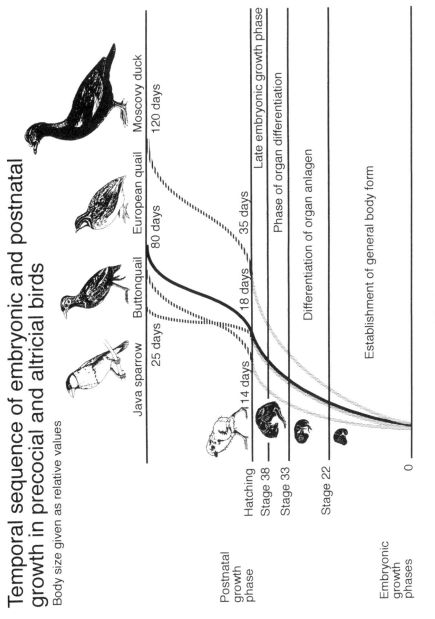

Fig. 1

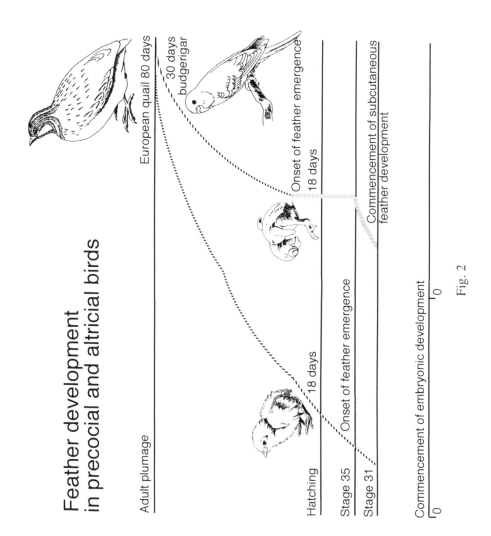

Fig. 2

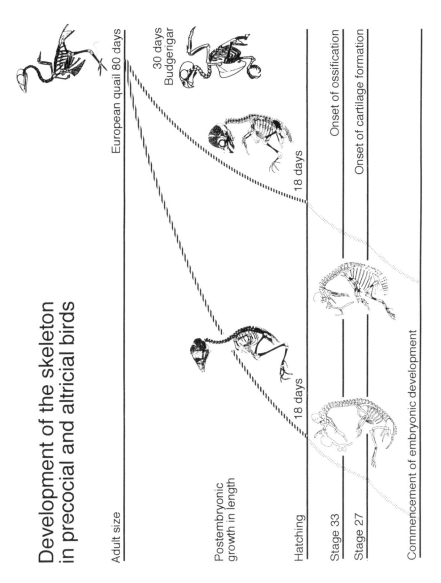

Fig. 3

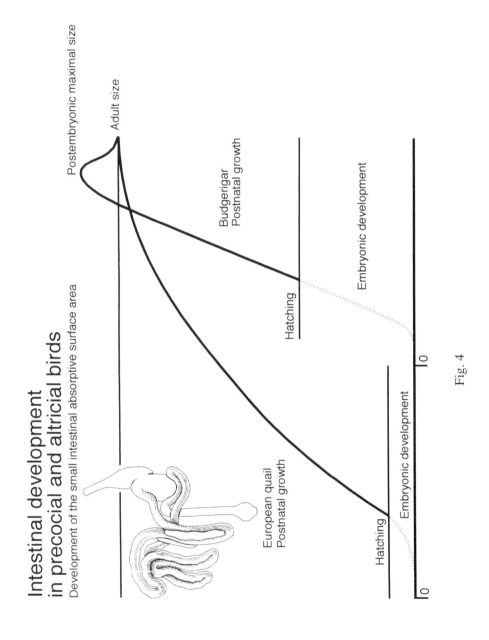

Intestinal development
in precocial and altricial birds
Development of the small intestinal absorptive surface area

Fig. 4

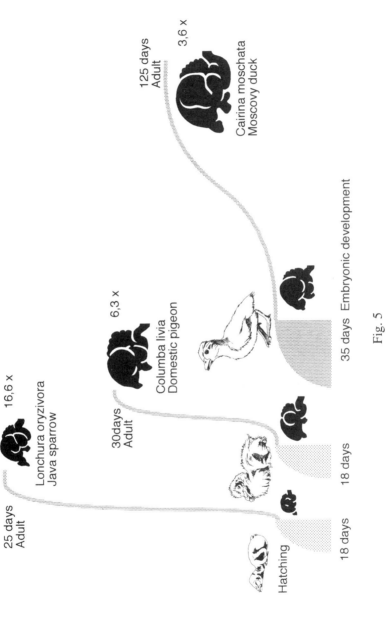

Brain growth after hatching in precocial and altricial birds

25 days Adult

16,6 x

Lonchura oryzivora
Java sparrow

30 days Adult

6,3 x

Columba livia
Domestic pigeon

125 days Adult

3,6 x

Cairina moschata
Moscovy duck

Hatching

18 days

18 days

35 days Embryonic development

Fig. 5

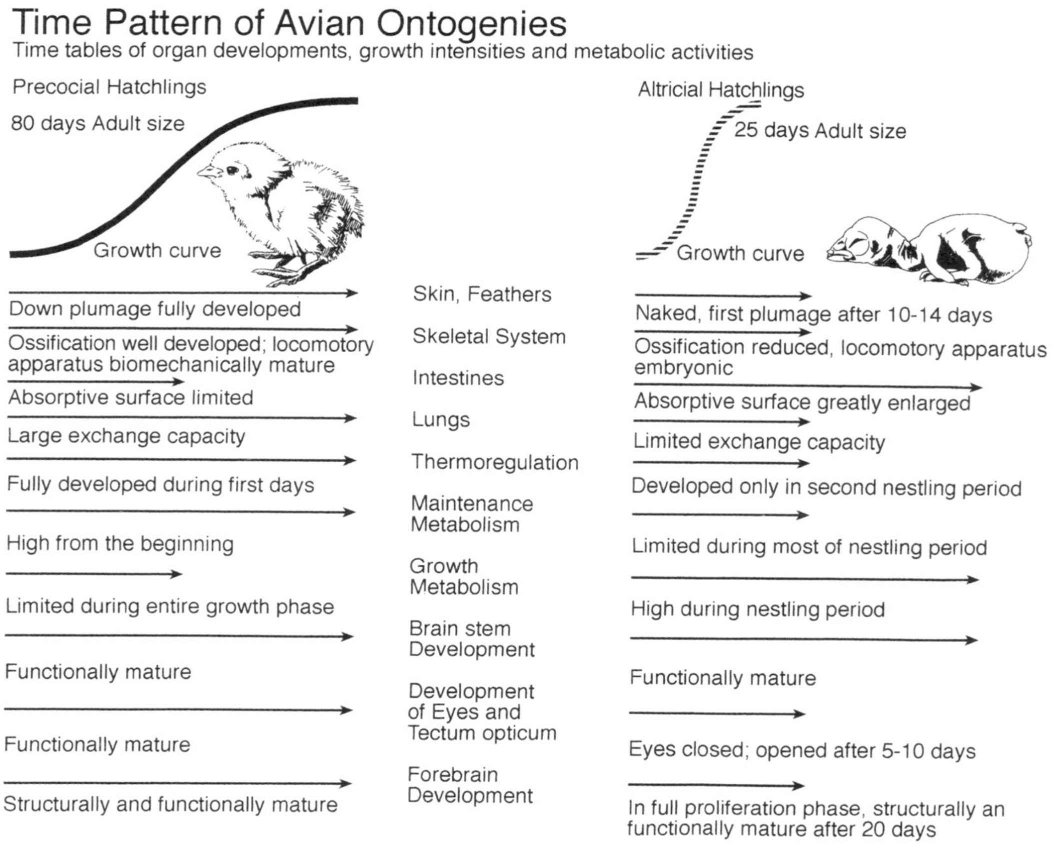

Time Pattern of Avian Ontogenies
Time tables of organ developments, growth intensities and metabolic activities

Precocial Hatchlings

80 days Adult size

Growth curve

Altricial Hatchlings

25 days Adult size

Growth curve

	Precocial	Altricial
Skin, Feathers	Down plumage fully developed	Naked, first plumage after 10-14 days
Skeletal System	Ossification well developed; locomotory apparatus biomechanically mature	Ossification reduced, locomotory apparatus embryonic
Intestines	Absorptive surface limited	Absorptive surface greatly enlarged
Lungs	Large exchange capacity	Limited exchange capacity
Thermoregulation	Fully developed during first days	Developed only in second nestling period
Maintenance Metabolism	High from the beginning	Limited during most of nestling period
Growth Metabolism	Limited during entire growth phase	High during nestling period
Brain stem Development	Functionally mature	Functionally mature
Development of Eyes and Tectum opticum	Functionally mature	Eyes closed; opened after 5-10 days
Forebrain Development	Structurally and functionally mature	In full proliferation phase, structurally an functionally mature after 20 days

Fig. 6

An outline of a theory of the constructional constraints governing early organismic evolution

Winfried Stefan Peters & Bernd Herkner

Introduction

The origin of life is one of the most fascinating biological enigmas (Dyson 1985, de Duve 1991). A multitude of hypotheses is available, ranging from the classical "primordial broth" (Haldane 1929) to the "genetic takeover" (Cairns-Smith 1982). Most of these models center on biochemical, metabolic, or genetic aspects. To our knowledge a comprehensive investigation into the mechanical constraints governing the early development of organismic constructions is not available to date. Attempting to close this gap, we here present an outline of a theory of the transformation of constructions in early evolution.

Theoretical Premises

In Bernal's now classic review "The Origin of Life" of 1967 a brief summary of the history of research into prebiotic evolution is given, which merits closer inspection. According to Bernal, life's origin is a "question that was evaded at the beginning of evolutionary theory because it was quite enough to have to fight for the animal ancestry of man, not to have to go on to explain the inorganic origin of life" (Bernal 1967, p.1). But the delayed interest in earliest life forms was allegedly not only due to philosophical and religious opposition against the notion of an inorganic origin of life, but also because biochemistry and molecular biology, "the sciences through which it could be understood, had not yet themselves developed" (ibid., p.2). However, this view of the two causes hampering progress in prebiotic evolution research, namely a general philosophical orientation adverse to evolutionary theory and a lack of appropriate methodologies, is probably not correct, and certainly not complete. We wish to give but one example for the doubtfulness of the historiographical reconstruction as such. The adsorption of simple organic compounds to mineral surfaces might have played a major role in chemical evolution, because the compounds were thus accumulated to the high concentrations necessary for the occurence of further reactions. This notion is commonly known today as the "Bernal hypothesis", after its assumedly first proponent (Bernal 1951). Ironically, very similar ideas indeed have been discussed in considerable detail by Chamberlin & Chamberlin in a widely ignored paper already in 1908.

However, more important than historiographical details is the question, whether the above interpretation actually addresses the central problem. Let us turn to Darwin's "Origin of Species". At first glance it appears surprising that Darwin very sparsly comments on the origin of life in this fundamental work. In the last edition revised by himself the crucial passage reads: "It can hardly be supposed that a false theory would explain, in so satisfactory a manner as does the theory of natural selection, the several large classes of facts above specified. ... It is no valid objection that science as yet throws no light on the far higher problem of the essence or origin of life. Who can explain what is the essence of the attraction of gravity?" (Darwin 1872, p. 421). A few pages later Darwin assumes that all living organisms "descended from some one prototype" (p. 424), but hastens to assure that "this inference is chiefly grounded on analogy, and it is immaterial whether or not it be accepted" (p. 425). Although Darwin speculated on life's beginning in a "warm little pond" (F. Darwin 1899, p. 17), he maintained that the origin of life was irrelevant to his theory, just as "the first origin of matter" was to "the laws of chemical attraction" (letter to J.D. Hooker, Burkhardt & Smith 1989, p. 350). While this line of argumentation certainly had the welcome effect of avoiding a foreseeable counterattack from the conservative and clerical establishment (Desmond & Moore 1991), it also possessed conceptual significance in itself. The idea of evolution by means of natural selection was not a theory of organisms, but one of their dynamics in time. It rested on general, yet unexplained organismic characters (variability, tendency to overproduce offspring), insofar as they bore on the historical development of organisms. These characters served as premises, not as explananda of the theory. Darwin obviously realized, broadly speaking, that a theory of the dynamic behavior of a class of natural objects, which was based on these very objects' general characters, neither needed to, nor could be expected to explain the coming into being of the particular class of objects. Thus, his self-restraint concerning the origin of life was based on conceptual grounds rather than on merely tactical considerations.

In the present essay we investigate constraints in early organismic constructions, using concepts borrowed from contemporary cell biology. Thus, our analysis will be subject to restrictions of a similar type as Darwin's theory of natural selection was; we will probably not be able to describe the transformation of "non-constructions" into constructions, as it were. Nevertheless, there obviously are interfaces between the reconstruction of early organismic constructions on one side, and for instance theories of chemical evolution on the other. But these interfaces are methodological in nature; temporal sequences in the sense of "first chemical evolution, followed by protocell evolution, afterwards organismic evolution", are not necessarily implied. For example, it obviously is a very interesting question, how heritability of functional characters was established. For our present purpose, however, it is entirely irrelevant whether the putative genetic material within the constructions we describe was DNA, RNA, proteins, clay minerals, or whatever. By thus restricting our theorizing to a narrow but well-defined range, we hope to obtain an unambiguous answer to a specific question.

The present paper links up with numerous studies, in which organismic constructions and their transformations were described in a terminology defined in analogy to the technical principles governing the action of man-made machines (e.g. W.F. Gutmann 1972, Peters 1985, Herkner 1989). This "machine analogy" (W.F.

Gutmann & Bonik 1981), which has been philosophically refined during the years (Weingarten 1995 and references therein), must not be misconstrued to mean that organisms are solely explanable as, or even *are* nothing but machines; it rather is shorthand for a methodological principle, which aims at avoiding circular definitions by relying on practical human experience in manipulating the outer world as a proto-biological standard of reference (W.F. Gutmann & Weingarten 1992; M. Gutmann 1993, 1995, this volume; Edlinger 1995; we here will refer to this approach as constructionalistic, since the direct English translation constructivistic is ambiguous). Applying the constructionalistic principle means to make biological objects accessible in a terminology developed in analogy to purposeful human activity. This approach should render the definitions of the terminology used reproducible. However, the "machine analogy" is only one out of several possible proto-biological standards. As any model it has its limits, and it obviously is not suitable for every purpose; suitability has to be proven in application. Thus a whole set of models might be required to approach what may be conceived as a single scientific problem, depending on their suitability regarding various puposes. For the sake of proper communication and reproducibility it is therefore necessary to always make explicit the standard of reference and the way of model constitution. We here use the machine-analogy in the explicated sense, since this appears the most suitable approach to constructional transformations.

In applying the machine-analogy a special problem arises in our particular case, since we deal with organismic conctructions of microscopic dimensions. There is no purposeful, direct human action in this realm, and neither exist man-made tools or machines the size of bacteria. On the other hand, D'Arcy Thompson (1966) was certainly correct in stressing that the behavior of objects with greatly different sizes is governed by different sets of physical rules. Hence, if we construct models of microscopic organisms, we cannot rely on classical newtonian or hydraulic mechanics; we have to depend on scientific experience as far as the manipulation of microscopic structures is concerned. Notably, the bulk of our insight into cellular energetics and the functional nature of biological membranes has been gained not from direct observation, but from experiments on isolated biological or artificial membranes with techniques that perplexingly closely resemble tests of electrical conductors or electronic circuits (for examples see Eisenberg & McLaughlin 1976; Sakmann & Neher 1983; Alvarez 1986; Nicholls & Ferguson 1992). We will draw on this type of practical experience in the following analysis.

Empirical Premises

Life in the biological sense does not exist independently of the distinct units we call organisms. The distinctiveness is maintained by the organisms' ability to create and control an inner milieu distinct from the outside world (an der Heiden, Roth & Schwegler 1985). In all known organisms the barrier which ultimately enables this ability is the cell membrane, a lipid bilayer enclosing the living protoplasma. Bilayer vesicles may form spontaneously in mixtures of lipids and water (Hargreaves,

Mulvihill & Deamer 1976; Israelachvili, Mitchell & Ninham 1977; Tanford 1978).
It appears widely accepted that the first organismic units distinguishable from their
environment arose by such spontaneous lipid vesicle formation (Deamer & Oro
1980; Koch 1985; Deamer 1986; Fleischaker 1990; Oro & Lazcano 1990; W.F.
Gutmann 1996). In our further considerations, we will stick to lipid bilayers as the
organism/environment boundary. Noteworthily this restriction is a self-imposed one
in the sense discussed above; we do not claim that other types of barrier are incon-
ceivable to have played a role in living structures. However, the physical properties
of lipid bilayers can be described, and the unavoidable consequences once closed
lipid vesicles exist can be evaluated.

Lipid bilayers form a dual phase transition between two aqueous compartments.
They are highly permeable to small hydrophobic compounds, and poorly permeable
to hydrophilic or charged molecules, or to big-sized ones in general (Deamer &
Bramhall 1986). Noteworthily, the permeability for water is orders of magnitude
greater than for ions (Jain 1972). Differential permeabilities together with electrostatic surface charges can give rise to electrical potential differences across the bilayer (Bangham, Standish & Watkins 1965; Jain 1972; Honig, Hubbell & Flewelling 1986; Pohorille & Wilson 1995). Contamination of the bilayer by hydrophobic
compounds might greatly change its characters (Grote, Syren & Fox 1978); enlight-
ening examples are given by Seta et al. (1985) and Steinberg-Yfrach et al. (1996),
who report the transduction of light energy into electrochemical proton gradients in
artificial membranes containing simple pigments.

Lipid bilayer vesicles establish chemically isolated and electrically insulated
"non-environment"-units, as it were. But it is equally important to note what they do
not accomplish. Lipid membranes have the character of fluids (Haydon & Taylor
1963; Jain 1972; Israelachvili 1991); they cannot withstand notable hydrostatic pressure gradients perpendicular to their surfaces, or serve as abutments where mechanical work is performed (as by the cytoskeleton in modern cells, which usually acts
on extracellular material; Ettinger & Doljanski 1992; Ingber et al. 1993). Let us
elaborate the consequences of this fact in the context of Darwin's famous "warm,
little pond". Imagine such a tidal pond at the sea shore undergoing an evaporative
volume loss of 75% in the sun, followed by rainfall. In this pond spontaneously
occuring lipid bilayer vesicles would first shrink to roughly a quarter of their original size due to osmotically driven water efflux. Then they would swell again in
proportion with the rain-induced dilution of the medium. If the diluted sea water
would reach 10% of the original osmolarity, the vesicles would have to expand to
the ten-fold of the original volume. Given a spherical shape, this would imply a 4.6-
fold increase of surface area, and a corresponding increase of lipid material incorporated into the bilayer. Since this appears unlikely to happen, one might suppose that
the decline in medium osmolarity could be counteracted by an increasing hydrostatic pressure in the vesicles. However, with an initial osmotic potential of about 2.5
MPa (as in "modern" sea-water) the hydrostatic pressure gradient required over the
bilayer would be roughly 2.2 MPa (i.e. the ten-fold of the gradient between a car
tire's filling and the atmosphere) in our example. Quite evidently, hydration-dehy-
dration cycles similar to the one described lead to a complete breakdown of vesicle

structure. If the process is reversible, an uptake of extravesicular substances might take place, which has been interpreted as a mechanism by which "protocellular structures" were produced under prebiotic conditions (Deamer & Barchfeld 1982). However, these structures would be destroyed during the hydration cycles at the same rate at which they were created. Therefore it appears unconceivable how stable protocells could develop from free-floating lipid bilayer vesicles. An analogous argument holds even in the case of an osmotically stable environment, because the establishment of a lipid bilayer vesicle is a step towards the beginning of organismic life only if it enables the occurence of an interior chemical composition distinct from the exterior one. But once chemical gradients are present, osmotic ones will develop due to the differential permeabilities of the lipid bilayer membrane. If we accept that organismic units became distinguishable from the outer world by a lipid membrane barrier surrounding them – and at some stage lipid membranes certainly did adopt this role – then we are forced to conclude that the structures from which all modern life descended must have possessed the ability to counteract hydrostatic and osmotic pressure gradients.

The above example describes exactly the situation met by modern organisms living in intertidal zones (Kirst 1985, 1989). Here the principal constraint in all organismic structures, namely the necessity to control its own osmotic properties, becomes visible in the extreme. This control over fluxes of water and permeable solutes in structures bounded by lipid membranes can only be exerted by two mechanisms. First, osmotically active substances can be activly transported over the membrane (as in typical animal cells). Second, the cellular structure might be embedded in a rigid extracellular matrix, which mechanically counteracts osmotically induced internal pressure (as in typical bacteria and plant cells).

Which of these mechanisms was employed by the first lipid membrane bounded structures? It is important to realize that active transport processes, when looked at as a whole, will always have to be energetically uphill in order to counteract passive water fluxes. Such processes do not only depend on structures which can accomplish solute transport against electrochemical gradients in a regulated manner, but also require a constant source of energy. Quite contrarily, the requirement for a rigid extracellular matrix is a very unspecific one, which might even be satisfied by inorganic structures such as porous minerals. If we accept Ockham's razor as a tool to establish plausibility arguments, and accordingly prefer the hypothesis based on the smallest number of ad hoc premises, we are forced to assume that the first membrane bounded structures stable enough to reach a state from which on organismic evolution could occur, were embedded in a pre-existent matrix.

In protocells of the type described osmotically induced pressure is counteracted by the rigid outer envelopment, which also determines the overall shape of the structure. It might seem hard to understand how such a construction could have transformed into one in which energy consuming membrane transport regulates internal osmolarity, and a cytoskeleton controls the outer shape; our protocell almost looks like a constructional dead end. However, the paradox is caused by an erroneous identification of structures with their functions. Both a complex cytoskeleton and sophisticated membrane transport systems are found in walled cells today (Harold

1986; Lloyd 1991). In contrast to locomotive animal cells, their role in cells with rigid walls is the control of intracellular structure and chemical homeostasis, respectively. In a similar way, a protocytoskeleton might have functioned primarily as a means to organize internal structure. We suppose that the earliest form of mechanical work probably was an ionic strength dependent conformation change within a gelatinous substance; the spasmonemes of certain protozoa provide a useful model for this mechanism (Amos 1971; Weis-Fogh and Amos 1972). On the other side, membrane transport mechanisms emerged as tools for the control of internal solute composition. The osmoregulating role of ion pumps, and the locomotive function of the cytoskeleton as found in modern animal cells thus are interpreted as secondary achievements in cellular evolution (compare Stein 1995).

The energetics of the protocell must be considered tightly coupled to its structure. The topologically closed lipid membrane creates the structural conditions under which electrochemical potential differences may develop. The structural requirements for the light-driven establishment of electrochemical proton gradients in lipid vesicles appear almost trivial if compared to the ad hoc postulates necessary to maintain the classical model of the emergence of fermentative organisms from the primordial "broth" (Morowitz 1992, Steinberg-Yfrach et al. 1996). Applying Ockham's razor once again, we have to assume for this reason alone that the first lipid membrane bounded protocells were light fueled. Light induced electrochemical gradients can serve as driving forces for secondary transport processes; this must not be confused with photosynthesis as such. Fortunately, the extant *Halobacterium halobium* renders a paradigmatic example for such a system (Wagner 1994).

Synthesis: A Scenario

The logical interrelations between the functional elements of the postulated protocell enforce a scenario of early cellular evolution, which contains surprisingly few degrees of freedom. In some crucial arguments it appears to link up neatly with the theory of surface catalized chemical evolution proposed by Wächtershäuser (1988). However, for reasons given above we restrict ourselves to the consideration of constructional aspects.

We postulate that films of hydrophobic molecules were adsorbed to a rigid surface, most likely on minerals such as clays or pyrite, which might have been mixed with simple organic molecules. Where this matrix formed three-dimensional networks, topologically closed lipid bounded vesicles developed. When pressure gradients over the lipid boundary occured due to differential solute permeabilities, the vesicles remained mechanically stable because of the external matrix. This was the essential structural pre-condition for the development of simple catalytic cycles, which eventually led to the establishment of primitive metabolic networks. The solute composition within the vesicle was probably kept very much different from that of the environment by lipophilic compounds forming membrane pores specific for particular ions; light dependent electrical potential differences then could stabilize ion distributions far from diffusive equilibrium. Some products of the primitive metabolic network might have been soluble under the unique conditons prevailing within the vesicles, but tended to polymerize or precipitate in the extravesicular milieu. If those compounds had a significant membrane permeability, they would have impregnated and reinforced the extracellular matrix. By this mechanism the initially inorganic matrix was stepwise replaced by an actively built organic one. As a result the protocells gained the option to conquer new environments apart from the material which had facilitated the origin of cellular structure.

Other products of the catalytic system might have coagulated in the unique ionic composition of the protocells, leading to an internal differentiation into gel and sol domains. Special physicochemical conditions prevailed in the immediate vicinity of the gelatinous substance due to surface charges, giving rise to specific interactions within the gel, and between the gel and freely diffusable molecules or other structures such as the lipid membrane. If the gel's conformation depended on the ionic strength within the vesicles, a primitive and initially unspecific mechanism for intravesicular translocation of gel-associated structures would have existed. Ionic strength dependent intravesicular contraction cycles would also have caused mechanical mixing of the vesicle's content, which would have provided an intravesicular transport mechanism in addition to diffusion. In parallel, the relatively unspecific pattern of selective membrane permeability evoked by lipophilic compounds in the lipid membrane developed into a more advanced machinery for chemical homeostasis in the vesicles. Active changes in intravesicular ion concentrations could have controlled the conformational status of the gelatinous substance. Thus the capacity to perform mechanical work was indirectly coupled to the primary energetic input (light) into the construction. The intravesicular gel turned stepwise into a protocytoskeleton.

At some stage active membrane transport and cytoskeletal activity had reached a level so advanced, that substantial defects in the synthesis of a functional cell wall ceased to be fatal; the cells had become able to control their internal hydrostatic pressure and outer shape by virtue of a highly efficient metabolic apparatus. Merging of cells became possible, which opened the options of phagocytosis and sexuality.

Under certain circumstances, e.g. when phagocytosis became obsolete after photoautotroph composite cells had been established by endosymbiosis, the secondary development of functional cell walls might have been favored because of the massive reduction in energy costs for pressure control in walled cells. We hypothesize that eucaryotic plants developed this way. In many organisms walled and wall-less phases alternate in ontogeny. Similarly, phylogenetic transformations from one cell type to the other probably occured repeatedly, depending on the particular circumstances. But while walled cells may survive phases of practically complete metabolic inactivity, wall-less constructions as such cannot exist without active osmoregulation, and a functional metabolic apparatus providing for the energetic demands of these regulatory processes. This insight creates formidable problems for all origin-of-life scenarios so far published. The present paper outlined one possibility to evade these problems.

Literature

Alvarez, O. (1986): How to set up a bilayer system. p. 115–130 in Miller, C. (ed.): Ion Channel Reconstitution. Plenum Press, New York.

Amos, W.B. (1971): Reversible mechanochemical cycle in the contraction of Vorticella. Nature **229**: 127–128.

an der Heiden, U., Roth, G. & Schwegler, H. (1985): Die Organisation der Organismen: Selbstherstellung und Selbsterhaltung. Functional Biology and Medicine **5**: 330–346.

Bangham, A.D., Standish, M.M. & Watkins, J.C. (1965): Diffusion of univalent ions across the lamellae of swollen phospholipids. Journal of Molecular Biology **13**: 238–252.

Bernal, J.D. (1951): The Physical Basis of Life. Routledge & Kegan Paul, London.

Bernal, J.D. (1967): The Origin of Life. The World Publishing Company, Cleveland.

Burkhardt, F. & Smith, S. (eds) (1989): The Correspondence of Charles Darwin, Vol. 5. Cambridge University Press, Cambridge.

Cairns-Smith, A.G. (1982): Genetic Takeover and the Mineral Origins of Life. Cambridge University Press, Cambridge.

Chamberlin, T.C. & Chamberlin, R.T. (1908): Early terrestrial conditions that may have favored organic synthesis. Science **28**: 897–911.

Darwin, C.R. (1872): The Origin of Species by Means of Natural Selection, or the Preservation of Favoured Races in the Struggle for Life. 6th ed., John Murray, London.

Darwin, F. (1899): Leben und Briefe von Charles Darwin, Vol. 3. Translated by J.V. Carus, 2nd ed., E. Schweizerbart'sche Verlagsbuchhandlung, Stuttgart.

Deamer, D.W. (1986): Role of amphiphilic compounds in the evolution of membrane structure on the early earth. Origins of Life **17**: 3–25.

Deamer, D.W. & Barchfeld, (G.L. 1982): Encapsulation of macromolecules by lipid vesicles under simulated prebiotic conditions. Journal of Molecular Evolution **18**: 203–206.

Deamer, D.W. & Bramhall, J. (1986): Permeability of lipid bilayers to water and ionic solutes. Chemistry and Physics of Lipids **40**: 167–188.

Deamer, D.W. & Oro, J. (1980): Role of lipids in prebiotic structures. BioSystems **12**: 167–175.

de Duve, C. (1991): Blueprint for a Cell: The Nature and Origin of Life. Neil Patterson Publishers/ Carolina Biological Supply Company, Burlington (NC).

Desmond, A. & Moore, J. (1991): Darwin. Michael Joseph Ltd.,London.

Dyson, F. 1985. Origins of Life. Cambridge University Press, Cambridge.

Edlinger, K. (1995): Elemente einer konstruktivistischen Begründung der Organismuslehre. Aufsätze und Reden der Senckenbergischen Naturforschenden Gesellschaft **43**: 87–104.

Eisenberg, M. & McLaughlin, S. (1976): Lipid bilayers as models of biological membranes. BioScience **26**: 436–443.

Ettinger, L. & Doljanski, F. (1992): On the generation of form by the continuous interactions between cells and their extracellular matrix. Biological Reviews **67**: 459–489.

Fleischaker, G.R. (1990): Origins of life: An operational definition. Origins of Life and Evolution of the Biosphere **20**: 127–137.

Grote, J.R., Syren, R.M. & Fox, S.W. (1978): Effect of products from heated amino acids on conductance in lipid bilayer membranes and nonaqueous solvents. BioSystems **10**: 287–292.

Gutmann, M. (1993): Der Vergleich als Konstruktion. Aufsätze und Reden der Senckenbergischen Naturforschenden Gesellschaft **40**: 45– 60.

Gutmann, M. (1995): Modelle als Mittel wissenschaftlicher Begriffsbildung: Systematische Vorschläge zum Verständnis von Funktion und Struktur. Aufsätze und Reden der Senckenbergischen Naturforschenden Gesellschaft **43**: 15–37.

Gutmann, W.F. (1972): Die Hydroskelett-Theorie. Aufsätze und Reden der Senckenbergischen Naturforschenden Gesellschaft **21**: 1–91.

Gutmann, W.F. (1996): Gibt es Alternativen für die Entwicklung der Organisation von Lebewesen? Natur und Museum **126**: 250–261 and 283–297.

Gutmann, W.F. & Bonik, K. (1981): Kritische Evolutionstheorie. Gerstenberg Verlag, Hildesheim.

Gutmann, W.F. & Weingarten, M. (1992): Maschinentheoretische Grundlagen der organismischen Konstruktionslehre. Philosophia Naturalis **28**: 231–256.

Hargreaves, W.R., Mulvihill, S.J. & Deamer, D.W. (1977): Synthesis of phospholipids and membranes in prebiotic conditions. Nature **266**: 78–80.

Haldane, J.B.S. (1929): The origin of life. The Rationalist Annual **148**: 3–10.

Harold, F.M. (1986): The Vital Force. W.H.Freeman, New York.

Haydon, D.A. & Taylor, J. (1963): The stability and properties of bimolecular lipid leaflets in aqueous solutions. Journal of Theoretical Biology **2**: 281–296.

Herkner, B. (1989): Die Entwicklung der saltatorischen Bipedie bei Säugetieren innerhalb der Tetrapodenevolution. Courier Forschungsinstitut Senckenberg **111**: 1–102..

Herkner,B. (1999): Über die evolutionäre Entstehung des tetrapoden Lokomotionsapparates der Landwirbeltiere. Ein konstruktionsmorphologisches Transformationsmodell auf evolutionstheoretischer Grundlage. carolinea, Beiheft 13. 353 S. Karlsruhe.

Honig, B.H., Hubbell, W.L. & Flewelling, R.F. (1986): Electrostatic interactions in membranes and proteins. Annual Reviews of Biophysics and Biophysical Chemistry **15**: 163–193.

Ingber, D., Karp, S., Plopper, G., Hansen, L. & Mooney, D. (1993): Mechanochemical transduction across extracellular matrix and through the cytoskeleton. p. 61–79 in Frangos, J.A. (ed.): Physical Forces and the Mammalian Cell. Academic Press, San Diego.

Israelachvili, J.N. (1991): Intermolecular and Surface Forces. 2nd ed., Academic Press, London.

Israelachvili, J.N., Mitchell, D.J. & Ninham, B.W. (1977): Theory of self-assembly of lipid bilayers and vesicles. Biochimica et Biophysica Acta **470**: 185–201.

Jain, M.K. (1972): The Bimolecular Lipid Membrane. Van Nostrand Reinhold Company, New York.

Kirst, G.O. (1985): Osmotische Adaptation bei Algen. Naturwissenschaften **72**: 125–132.

Kirst, G.O. (1989): Salinity tolerance of eukaryotic marine algae. Annual Reviews of Plant Physiology and Plant Molecular Biology **40**: 21–53.

Koch, A.L. (1985): Primeval cells: Possible energy-generating and cell division mechanisms. Journal of Molecular Evolution **21**: 270–277.

Lloyd, C.W. (ed.) (1991): The Cytoskeletal Basis of Plant Growth and Form. Academic Press, London.

Morowitz, H.J. (1992): Beginnings of Cellular Life. Metabolism recapitulates Biogenesis. Yale University Press, New Haven.

Nicholls, D.G. & Ferguson, S.J. (1992): Bioenergetics 2. Academic Press, London.

Oro, J. & Lazcano, A. (1990): A holistic precellular organization model. p. 11–34 in Ponnamperuma, C. & Eirich, F.R. (eds.): Prebiological Self-Organization of Matter. A. Deepak Publishing, Hampton.

Peters, D.S. (1985): Mechanical constraints canalizing the evolutionary transformation of tetrapod limbs. Acta Biotheoretica **34**: 157–164.

Pohorille, A. & Wilson, M.A. (1995): Molecular dynamics studies of simple membrane-water interfaces: Structure and functions in the beginnings of cellular life. Origins of Life and Evolution of the Biosphere **25**: 21–46.

Sakmann, B. & Neher, E. (eds.). (1983): Single Channel Recording. Plenum Press, New York.

Seta, P., Bienvenue, E., Moore, A.L., Mathis, P., Bensasson, R.V., Liddell, P., Pessiki, P.J., Joy, A., Moore, T.A. & Gust, D. (1985): Photodriven transmembrane charge seperation and electron transfer by a carotenoporphyrinquinon triad. Nature **316**: 653–655.

Stein, W. (1995): The sodium pump in the evolution of animal cells. Philosophical Transactions of the Royal Society, London, Series B **349**: 263–269.

Steinberg-Yfrach, G., Liddell, P.A., Hung, S.-C., Moore, A.L., Gust, D. & Moore, T.A. (1996): Conversion of light energy to proton potential in liposomes by artificial photosynthetic reaction centres. Nature **385**: 239–241.

Tanford, C. (1978): The hydrophobic effect and the organization of living matter. Science **200**: 1012–1018.

Thompson, D.W. (1966): On Growth and Form. (abbr. ed. by J.T.Bonner) Cambridge University Press, Cambridge.

Wagner, G. (1994): Halobakterien – Leben im biotischen Grenzbereich. p. 141–157 in Hausmann, K. & Kremer, B.P. (eds.): Extremophile. VCH Verlagsgesellschaft, Weinheim.

Wächtershäuser, G. (1988): Before enzymes and templates: Theory of surface metabolism. Microbiological Reviews **52**: 452–484.

Weingarten, M. (1995): Einleitung: Form, Struktur und Funktion. Aufsätze und Reden der Senckenbergischen Naturforschenden Gesellschaft **43**: 7–14.

Weis-Fogh, T. & Amos, W.B. (1972): Evidence for a new mechanism of cell motility. Nature **236**: 301–304.

Monophyly of Metazoa:
Phylogenetic analyses of genes encoding Ser/Thr-kinases and a receptor Tyr-kinase from Porifera [sponges]

Werner E.G. Müller, Isabel M. Müller, Vera Gamulin, Vadim Kavsan,
Michael Kruse and Alexander Skorokhod

1. Introduction

Protein tyrosine kinases [Tyr-kinases] [PTKs] represent a large group of enzymes that specifically phosphorylate tyrosine residues (Hardie and Hanks 1995). They play important roles in the response of cells to different extracellular stimuli and are essential proteins most notable for control of growth and differentiation (Hunter et al. 1992). Many PTKs serve as receptors and signal transducers for circulating peptide hormones and growth factors. PTKs were first discovered in oncogenic retroviruses and subsequently identified and analyzed in a variety of different metazoan organisms (Hanks and Quinn 1991). Hundreds of PTK primary structures are known; all have been isolated from metazoan organisms (Hardie and Hanks 1995). PTKs together with serine/threonine [Ser/Thr]-kinases represent the largest known protein superfamily (Hardie and Hanks 1995).

All PTKs possess a closely related Tyr-kinase [TK]-domain which is specific for the phosphorylation of tyrosine only (Hunter et al. 1992; Ullrich and Schlessinger 1990). It was estimated by Hunter (1987) that the number of protein kinases, including TKs, present in a highly evolved metazoan organism might be one thousand or even more (Hunter 1987). This number is close to our present knowledge. It can be expected that many of these enzymes, especially if they are members of the same [or closely related] subfamilies have very probably only a recent biological history. PTKs are divided into two major groups, the receptor Tyr-kinases [RTKs], which are membrane spanning molecules with similar overall structural topologies, and the non-receptor TKs, also composed of structurally similar molecules. Members of the PTKs are further classified on the basis of their structural [functional] similarities, and divided into more than twenty subfamilies (Hardie and Hanks 1995). Phylogenetic analysis, based on the alignment of TK-domain amino acid [aa] sequences, indicates that all TK proteins have one common ancestor (Hardie and Hanks 1995). Since PTKs were found only in the metazoan [multicellular] organisms it is reasonable to postulate that these enzymes derive from one common ancestral molecule (Müller 1995).

2. Monophyly of Metazoa

Evolution is a gradual process whereby new genes are primarily formed by either gene duplication or exon shuffling. However, new proteins can also be produced by overlapping genes, alternative splicing or gene sharing. These facts imply *(i)* that proteins found for the first time in a given phylum contain elements, modules, which are present already in ancestral protein(s) of members of phylogenetically older phyla and *(ii)* that new combinations of modules create proteins that possess new functions.

Therefore, we postulate that animals – like sponges – which are positioned at the border between Protoctista and Metazoa are especially rich in molecules having already the functions of those in higher Metazoa as well as in ancestral molecules comprising modules for structural and functional proteins found in higher phyla. The structures of the characteristic metazoan genes and proteins required for *(i)* tissue formation [galectin, collagen, integrin], *(ii)* signal transduction [receptor Tyr-kinase], *(iii)* transcription [homeodomain- and MADS-box-containing proteins (Müller 1997a)], *(iv)* immune reaction [heat shock proteins, proteasome, proteins featuring scavenger receptor cysteine-rich domains (Müller 1997b)] and *(v)* sensory tissue [crystallin] have been identified in *Geodia cydonium* (Fig. 1) and found to display high similarity to molecules from members of higher metazoan phyla (Müller 1997a).

Based on these sequence data it is reasonable to adopt the view that Porifera should be placed into the kingdom Animalia together with the Metazoa (Müller et al. 1994; Müller 1995; Müller 1997a). In addition, as taken from the first sponge genes, especially that coding for RTK, it is now established that modular proteins, composed by exon-shuffling, are common to **all** metazoan phyla; a detailed description is given elsewhere (Müller and Müller 1997). This mechanism of exon-shuffling is apparently absent in plants and protists (Patthy 1995). If this view can be accepted then the "burst of evolutionary creativity" (Patthy 1995) during the period of Cambrian explosion which resulted in the big bang of metazoan radiation (Lipps et al. 1992) was driven by the process of modularization. During this process the already existing domains were transformed into mobile modules allowing the composition of mosaic proteins; Fig. 1.

Most cDNAs coding for proteins that are discussed below and have been used to establish *(i)* the classification within the Porifera in particular, and *(ii)* the monophyly of Metazoa in general, come from the Demospongiae *Geodia cydonium* (Jameson) and *Suberites domuncula* (Olivi), the Calcarea *Sycon raphanus* (Schmidt) and the Hexactinellida *Rhabdocalyptus dawsoni* (Lambe).

3. Cloning of the ancient receptor Tyr-kinase gene in *G. cydonium*

We used a *G. cydonium* RTK cDNA probe (Schäcke et al. 1994c) to screen a genomic DNA library. The genomic DNA clone containing the entire coding sequence of sponge RTK was analyzed. It contains the gene for RTK, designated here RT-

Kinase_gene of 4,871 bp (Fig. 2 A; Gamulin et al. 1997). A comparison – of the corresponding segments – of the nt as well as the deduced aa sequences of both the gene and the cDNA encoding the RTK revealed that they differ only slightly in the respective nucleotide (nt) as well as in the deduced amino acid (aa) sequence. The size of the sponge RTK mRNA was identified by Northern blotting at 3.3 kbp. The deduced aa sequence of the exons of RT-Kinase_gene reveals 917 aa indicating an M_r of 101,309 [corresponding to 2.8 kbp].

3.1. Gene structure of sponge RTK

The coding sequence of RTK contains three exons. The putative structure of the sponge RTK gene, RT-Kinase_gene, shows *(i)* the extracellular part, comprising the Pro-Ser-Thr[P/S/T]-rich domain and two complete immunoglobulin [Ig]-like domains (Schäcke et al. 1994a; Schäcke et al. 1994b), *(ii)* the transmembrane domain, *(iii)* the juxtamembrane region and *(iv)* the catalytic TK-domain (Fig. 2 A). Two introns have been found, the first is located between the two Ig-like domains and the second intron between the second Ig-like domain and the transmembrane region in the extracellular part of RT-Kinase_gene. The introns are of medium size [≈180 bp and 508 bp, respectively]. However, the rest of the gene, comprising the transmembrane domain, the juxtamembrane region and the catalytic TK-domain is coded by one single exon.

3.2. Tyr-kinase catalytic domain

A homology search with the TK-domain aa sequence, deduced from the gene sequence, was performed. All 50 most homologous sequences to sponge RTK contained the TK-domains of PTKs. Many of those PTKs were orthologous gene products from different organisms. One representative specimen from each group of closely related enzymes was used for the multiple sequence alignment. The alignment was computed and subsequently manually adjusted to delineate the TK subdomains I-XI (Gamulin et al. 1997). It was found that the sponge TK-domain contains all conserved aa, or streches of aa, known to be important for the function of these enzymes (Geer et al. 1994).

3.3. Phylogenetic tree

Highest homology (45%) in the *G. cydonium* TK-domain within the subfamily of RTK was found with genes coding for the nerve growth factor receptors [*TRKB*_Mouse; *TRKC*_Pig and *TRKA*_Human]. These TKs contain two Ig-like domains in the extracellular part of the proteins as well. RTK genes of the insulin receptor subfamily [*IRR*_Human; *IG1R*_Human; *INSR*_Human] and of related enzymes [*ROS*_Human; *SEV*_Dmela] also show significant homology in the deduced

TK-domain to the sponge RT-Kinase_gene. Lower homologies in the range of 32% were found with the fibroblast growth factor subfamily of RTKs [*FGR1*_Human], having three Ig-like domains in the extracellular part, and the *TIE* receptor PTK [*TIE*_Mouse; two Ig-like domains]. Interestingly, several non-receptor TKs were among the 50 proteins having highest homology in the TK-domain with the sponge RTK. The TK-domains of non-receptor TK genes from the Abl subfamily [*ABL*_Dmela; *ABL*_Human] show remarkable homology (41%) and other selected non-receptor TKs [*TEC*_Dmela; *FER*_Human and *FES*_Human] still have a homology of over 35%.

The dendrogram of the alignment of TK-domains of 15 PTKs is shown in Fig. 3 A. All RTKs used for the dendrogram fall in one branch of the tree while all non-receptor TKs are grouped in a second one; sponge RTK [GCTK_Geodia] is placed in a separate branch, which splits off first from the common tree of metazoan PTKs.

3.4. Promotor of RT-Kinase_gene

The potential promotor elements, TATA box [nt -625 – -620], GC-box [nt -648 as a center] and Cap signals [nt -580, – 599 and -602 as centers] are found. In addition, a CArG box-like sequence, CCTATATGG [nt 97 – 105], is present in the potential promotor region, which is known to be the target for the serum response factor, a transcription factor which is involved in the RTK signaling pathway (Shore and Sharrocks 1995); Fig. 4 C and Fig. 5. In addition, three further transcription binding sites are found in the RTK promotor, *(i)* elements for binding of the heat shock factors [HSFs], heat shock elements [HSEs], *(ii)* binding sites for homeodomains and *(iii)* the CF2-II binding site, controlling e.g. in *Drosophila melanogaster* the chorion development (Fig. 5).

3.4.1. Serum response factor

Stimulation of receptor Tyr-kinases, e.g. the epidermal growth factor tyrosine-kinase receptor, causes the expression of a number of genes without the need for prior protein synthesis (reviewed in: Greenberg and Ziff 1984). The rate of transcription of those immediate-early genes, with the proto-oncogene c-*fos* as a prominent member, increases – e.g. in fibroblasts – within 5 min by 50-fold and returns to a basal level after 30 min. After stimulation of a cell in response to binding of a growth factor RTK to its cognate ligand, a series of phosphorylation events are triggered, among them the activation of the mitogen-activated protein [MAP] kinase cascade which in turn results in the phosphorylation of the ternary complex factors [TCFs] and the ribosomal S6 protein kinase pp90^{rsk} (Sun and Tonks 1994). The pp90^{rsk} kinase itself phosphorylates the serum response factor [SRF] (Sun and Tonks 1994); Fig. 4 A. After phosphorylation, the SRF dimerizes and displays high affinity to the serum response element [SRE] within the c-*fos* promoter region where it controls together with TCFs, c-*fos* expression (reviewed in: Sun and Tonks 1994).

A cDNA clone encoding the sequence-specific DNA binding protein, SRF, has been isolated and analyzed from *G. cydonium* (Scheffer et al. 1997). Sequence comparisons revealed that the aa sequence shares high homology to human and *Xenopus laevis* SRFs. Alike these vertebrate sequences the sponge sequence displays the MADS-box motif in the DNA-binding domain (Fig. 4 B). The SRF, and also other transcription factors, such as ternary complex factors, recognizes the serum response element [CArG box] in promoter regions of a series of cellular immediate-early genes, whose expression is controlled by growth factors. A related CArG box sequence is present in the sponge gene, encoding the receptor Tyr-kinase (Gamulin et al. 1997); Fig. 4 C.

3.4.2. Heat shock factor

Heat shock factors [HSFs] recognize specific elements, heat shock elements [HSEs], in the promotors of heat shock protein genes (Morimoto 1993). HSEs are characterized by a palindromic, 14 bp heat shock consensus element, C--GAA-TTC---G (Bienz and Pelham 1987). In the putative promotor of the sponge gene three HSE sites with 100% identity are present (Fig. 5).

The HSF is activated by temperature, a process which occurs in two steps (Larson et al. 1988). First, heat shock protein [HSP] is activated to a form which binds to DNA by an ATP-dependent mechanism. Then a dimerization of the factor from the monomeric to the trimeric form takes place. At present, first data about the activation of genes involved during initiation of apoptosis are available in *G. cydonium*. However, it is not known if the induction of these genes involves a HSE/HSF interaction.

3.4.3. Homeodomain

Interesting is the finding that the promotor of the RTK shows also elements which resemble those recognized by homeodomains. Genes encoding homeodomains contain the homeoboxes. Several families of homeobox genes have recently been found in sponges (Seimiya et al. 1997).

4. Partial cloning of the Ser/Thr-kinase genes of *G. cydonium*

4.1. cDNAs encoding Ser/Thr-kinases

Ser/Thr-kinases are ubiquitously present in animal tissues; they respond to second messengers, e.g. Ca^{2+} and/or diacylglycerol, to express their activities. The first sponge cDNAs encoding Ser/Thr-kinases have been isolated from *G. cydonium*. One such Ser/Thr-kinease from *G. cydonium*, belongs to the "novel" [Ca^{2+}-independent] protein kinase C [PKC] – abbreviated nPKC – subfamily while the second one has the hallmarks of the "conventional" [Ca^{2+}-dependent] PKC [cPKC] subfamily (Kruse et al. 1996).

4.1.1. nPKC

The deduced aa sequence of nPKC (a scheme is given in Fig. 2 B) displays the following boxes characteristic for nPKC; the Ser/Thr-kinase catalytic domain, the putative ATP-binding domain, the pseudosubstrate segment and the phorbol esters/diacylglycerol binding domain as well as two zinc fingers (Kruse et al. 1996).

4.1.2. cPKC

The following units are present in the second sponge Ser/Thr-kinase, the cPKC; the pseudosubstrate segment, the "Greek key" motif, the typical phorbol esters/diacylglycerol binding domain, two zinc fingers, an ATP/GTP-binding site motif, the C2 domain signature, which is likely involved in Ca^{2+} binding and the Ser/Thr-kinase catalytic domain (scheme Fig. 2 C). Since this kinase has a C2 domain, it belongs to the "conventional" [Ca^{2+}-dependent] protein kinase C [cPKC] subfamily (Kruse et al. 1996).

4.2. Partial genes

Partial sequences of the genes from the nPKC as well as the cPKC from *G. cydonium* have been isolated (to be published).

4.2.1. nPKC

In a first approach we have sequenced the ATP-S/STK segment of the gene encoding the nPKC [nS/T-Kinase_gene] (Fig. 2 B) in oder to compare the intron distribution. In contrast to the RTK gene four introns have been found in the kinase domain (Fig. 2 B and Fig. 6). Interestingly, the first two introns, [intron a: 308 nts and intron b: 231 nts] flank the ATP binding site while intron c [157 nts] terminates – again important – the kinase domain; intron d [> 2.9 kbp] is positioned outside the kinase region.

4.2.2. cPKC

Only a small part of the cPKC gene [cS/T-Kinase_gene] is yet known. However, again – now becoming familiar – the introns frame the segments of the pseudosubstrate region and the first zinc finger (Fig. 2 C).

5. cDNAs encoding related Ser/Thr-kinases from sponges: Origin of Tyr-kinases

All of the known Ser/Thr-kinases and Tyr-kinases [both PTKs and RTKs] share a related catalytic domain of approximately 270 aa (Hanks, Quinn and Hunter 1988), which is further divided into 12 smaller subdomains. In order to investigate the phylogenetic relationships of the Ser/Thr-kinases and Tyr-kinases we have isolated them from different sponge species.

The phylum Porifera [sponges] is divided into *(i)* Demospongiae, *(ii)* Calcarea, and *(iii)* Hexactinellida. Member(s) of Demospongiae [*G. cydonium* and *S. domuncula*] and Calcarea [*S. raphanus*] have been used for our studies. cDNAs of cPKC and nPKC have been identified from these three species. In addition, the cDNA of a PKC-related kinase [PRK] (Palmer, Ridden and Parker 1995) was isolated from *S. domuncula*, and another one from the same species, belonging to the group of stress-responsive protein kinases [KRS] (Creasy and Chernoff 1995) has been cloned (Kruse et al. 1998).

An unrooted neighbour-joining tree was constructed from the the catalytic domains of the sponge cPKC, nPKC, KRS and PRK. It shows that the nPKC and the cPKC cluster together, while the third branch of the tree includes the KRS and the RTK (Fig. 3 B). The protein-kinase-C-related kinase from *S. domuncula* [PRK_SD] does not fall in any cluster. In the construction of the tree also the sequences of four different *src*-Tyr-kinases from *Spongilla lacustris* (Ottilie et al. 1992) habe been included. They branch off later than the *G. cydonium* RTK. The robustness of the inferred phylogenies was tested by bootstrap test and revealed that the branching of the nPKC and cPKC has a significance of >87%. The cluster, including the stress-responsive protein kinase [KRS_SD] and the RTK from *G. cydonium* is separated from the other two by a 100% significance. Within the latter branch the KRS diverged first, while the RTK from *G. cydonium* appeared later together with the *S. lacustris* src-sequence. Also a Ser/Thr and Tyr phosphorylating kinase from *Saccharomyces cerevisia* – SPK1_YEAST has been included into the alignment. The catalytic domain of this kinase is equally distantly related to the metazoan sponge PKCs as to the Tyr-kinases.

We conclude that – based on the deduced kinase data – the class Calcarea appeared earlier during evolution than the class of Demospongiae. Lastly, the phylogenetic analyses strongly suggest that the PTK have a common ancestor with the PS/TK superfamily from which the sponge sequences RTK_GC diverged first.

6. Relationship of sponge classes with respect to higher Metazoa: Isolation of a cDNA encoding a Ser/Thr-kinase from an Hexactinellida

6.1. Phylogenetic position of the classes of Porifera

Two alternative hypotheses have been proposed to explain the relationships between the major sponge classes. One groups the Porifera into the adelphotaxa Hexactinellida and Demospongiae/Calcarea based on the gross difference in tissue structure and on differences in the structure of the flagella, whose beating generates the feeding current through sponges (Mehl and Reiswig 1991). The other hypothesis assumes that the Demospongiae are more closely related to Hexactinellida based on presumed larval similarities (Böger 1988).

6.2. Relationship of sponge classes

In order to approach this question the cDNA encoding a protein kinase C belonging to the C subfamily [cPKC] has been isolated and characterised from the hexactinellid sponge *Rhabdocalyptus dawsoni* (Kruse et al. 1998). The two conserved regions, the regulatory part with the pseudosubstrate site, the two zinc fingers and the C2 domain, as well as the catalytic domain were used for phylogenetic analyses. Sequence alignment and construction of a phylogenetic tree (Fig. 3 C) from the catalytic domains revealed that the hexactinellid *R. dawsoni* branches off first among the metazoan sequences; the other two classes of the Porifera, Calcarea [the sequence from *S. raphanus* was used] and Demospongiae [sequences from *G. cydonium* and *S. domuncula* were used] branch off later. The statistically robust tree also shows that the cPKC sequences from higher invertebrates, *(i)* the pseudocoelomata [*C. elegans*], *(ii)* the protostoma *D. melanogaster* and *Aplysia californica* and *(iii)* the deuterostoma *Lytechinus pictus* and *X. laevis* are closest related to the calcareous sponge. This finding was also confirmed by comparing the regulatory part of the kinase gene. From this and related analyses of the 70 kDa heat shock protein isolated from the same sponge species (Koziol et al. 1997) it is justified to conclude *(i)* that within the phylum Porifera, the class Hexactinellida diverged first from a common ancestor to the Calcarea and the Demospongiae, which both appeared later, and *(ii)* that the higher invertebrates are closer related to the calcareous sponges (Müller et al. 1998).

This conclusion supports the separation of the Porifera into two subphyla (Reiswig and Mackie 1983) which was primarily based on the difference between syncytial and cellular tissues. Since the Hexactinellida are syncytial animals (Mackie and Singla 1983) and since the present analysis indicates that they branched off from a common ancestor earlier than other sponges, this could imply that multicellularity came about by division of a multinucleate syncytium rather than by aggregation of formerly single cells. However, it is also possible that the syncytial nature of the Hexactinellida reflects a reduction of a previously multicellular stage by fusion to form a syncytium.

7. Introns early or late?

Most of the informations about the PTKs in different organisms were obtained from studies of the corresponding cDNAs. During the last years the organization of a growing number of genes encoding these proteins has been analyzed. To the best of our knowledge, all studied genes contain introns in their TK-domain. Some of the genes encoding RTKs are over 100 kbp long with more than 20 introns, as in the case of insulin receptor subfamily of RTKs (Seino et al. 1989). The insulin receptor gene contains five introns in the TK-domain. Comparison of the exon structure of the TK-domains of this and three other TK genes [*ROS*, *SRC* and *ERBB2*] revealed that the exon-intron organization of this region has not been well conserved during evolution (Seino et al. 1989). According to the exon theory of genes (Gilbert et al.

1986; Gilbert 1987) it was proposed by Seino et al. (1989) that the putative ancestral TK-domain may have been assembled from 13 exons; consequently, introns must have been lost in a more or less random fashion from individual genes. Genes encoding human RTKs with three, five or seven Ig-like domains and the TK insert within the catalytic domain were recently studied in detail (Agnès et a. 1994; Rousset et al. 1995). In these reports introns were found at conserved positions within the TK-domains of these genes. These data were used to establish the phylogenetic relationships between three related subfamilies of RTKs, and it was proposed that all genes most probably evolved from the common ancestor already "in pieces" by successive duplications involving entire genes. However, the time when the first duplication might have occurred was not discussed.

The Tyr-kinase domain of the ancient enzyme, the RTK from the marine sponge *G. cydonium*, is – as mentioned – encoded by one single exon. There are many reasons to assume that *G. cydonium* RTK is the most ancient RTK analyzed so far (Schäcke et al. 1994a; Schäcke et al. 1994b; Schäcke et al. 1994c) which branched off first from the common tree of metazoan PTKs (Fig. 3 A). PTKs are found only in metazoan [multicellular] organisms and *G. cydonium* belongs to the oldest and the most simple metazoan phylum, the Porifera.

8. Evolution of Metazoan Genes

Based on sequence data from phyla higher than Cnidaria it was deduced that mosaic proteins are constructed after modularization of protein domains by intron insertions and subsequent dispersal through exon-shuffling (Patthy 1994). This constructive view implies that the ancestor molecules of Metazoa were without introns.

An experimental evidene for this assumption was lacking until the gene coding for the receptor Tyr-kinase from the sponge *G. cydonium* was cloned (Gamulin et al. 1997). Therefore, this gene can be considered to be a prototype of a metazoan mosaic protein. Cloning of the gene coding for the sponge RTK revealed – as stated – that it contains only two introns which are both present in the extracellular part of the protein (Gamulin et al. 1997); Fig. 2 A. All other RTK genes studied contain introns in their TK-domain as well. Some of the genes encoding RTKs are over 100 kbp long with more than 20 introns, as in the case of the insulin receptor subfamily of RTKs (Seino et al. 1989).

The ancestor of the RTK gene of *G. cydonium* is – as mentioned above – close to the Porifera. Evidence from sequence comparisons (see above) suggests that the kinase domain of RTK orginated from the ancestral, enzymic domain related to today's KRS Ser/Thr-kinase, a Ser/Thr-kinase. It is postulated that the coding exon for the kinase site in the *donor molecule* is bordered by introns (Fig. 7 phase I and II). Modularization [phase I] of a postulated donor protein d-P occurs by intron insertion – here the KRS-like kinase – under formation of a protomodule [phase II] (Patthy 1994) "Ex-2". In phase III the protomodule undergoes duplication forming "Ex-2,1" and "Ex-2,2". Module "Ex-2,2" is excised in phase IV. In phase V the excised "Ex-2,2" the former S/TK domain [KRS-like domain] is inserted into the *acceptor molecule*, the precursor gene RTK [p-RTK], under formation of a kinase domain now provided with an enzymic specificity for the aa Tyr. This ancestor molecule gave rise to the nowadays existing Tyr-kinases. A Ser/Thr-kinase which comprises two kinase domains [the ribosomal protein S6 kinase II] is known from vertebrates (Hardie and Hanks 1995). A schematic representation of the possible formation of the Ig-like module and its subsequent insertion into the RTK precursor gene is shown in Fig. 7.

Conclusion

Until recently, sponges have been regarded as colonies of unspecialized cells – "individual flagellates" – which require only cells that secrete adhesive glycoprotein and bind to it (Loomis 1988). This view has completely changed with the application of modern molecular biological techniques. The results – accumulated just during the past three years – reveal that sponges are composed of molecules which are typical for Metazoa. It appears that *all* basic molecules required for the maintenance of an individuum known from higher metazoan phyla are present already in sponges.

In a further step the question for the structure of the common ancestor gene for characteristic metazoan proteins – here the example TK-domains of PTK, including the *G. cydonium* RTK – must be answered. According to the exon theory of genes also called the "intron early" view (Gilbert et al. 1986; Gilbert 1987; Darnell 1978; Doolittle 1978), this common ancestor gene was present already in pieces. If this assumption should be correct, then (at least) in the sponge *G. cydonium* all introns have been eliminated from the TK-domain during the long period of the sponge's separate evolution. This view cannot be excluded considering the fact that – based on populational genetic studies – the sponge genome is particularly dynamic (Solé-Cava and Thorpe 1991). However, it is also – perhaps more – likely that introns, found in the TK-domain of existing PTKs genes were introduced after splitting off the sponge line from the common, primitive ancestral metazoan organisms. Recent investigations from the early supporters of "genes in pieces" and "intron early" view, speak against their own theory (Stolzfus et al. 1994). The most logical way to explain the results obtained from the analysis of the ancient *G. cydonium* RTK gene is to accept the "introns late" hypothesis (Orgel and Crick 1980; Cavalier-Smith 1991). Introns in the TK-domains of the TK proteins were introduced gradually during the [recent] evolution of these enzymes in the kingdom of Metazoa.

As a result it can be deduced that one major event which allowed evolution of Protozoa to Metazoa was the invention of the genetic apparatus to insert introns and hence to facilitate the creation of mobile modules (Patthy 1995), thus facilitating the *subsequent* process of exon-shuffling (Gilbert 1978).

With the isolation of the first sponge genes, especially that coding for RTK, it is now established that modular proteins, composed by exon-shuffling, are common to *all* metazoan phyla including sponges. This mechanism of exon-shuffling is apparently absent in plants and protists (Patthy 1995). If this view can be accepted then the "burst of evolutionary creativity" (Patthy 1995) during the period of Cambrian explosion which resulted in the big bang of metazoan radiation was driven by the process of modularization. During this process the already existing domains were transformed into mobile modules allowing the composition of mosaic proteins (Patthy 1994); Fig. 1. The use of modules, building blocks, for the creation of mosaic proteins became only possible after a new step in evolution was acquired – the invention of introns – which allowed exon-shuffling.

Acknowledgements
This work was supported by grants from the Stiftung Volkswagenwerk and the International Human Frontier Science Program [RG-333/96-M].

Literature

Agnès, F., Shamoon, B., Dina, C., Rosnet, O., Birnbaum, D. & Galibert, F. (1994): Genomic structure of the downstream part of the human FLT3 gene: exon/intron conservation among genes encoding receptor tyrosine kinase (RTK) of subclass III. Gene **145**:283–288.

Bienz, M. & Pelham, H.R.B. (1987): Mechanisms of heat shock gene activation in higher eukaryotes. Adv Genet **24**:31–72.

Böger, H. (1988): Versuch über das phylogenetische System der Porifera. Meyniana **40**:143–154.

Cavalier-Smith, T. (1991): Intron phylogeny: a new hypothesis. Trends Gen **7**:145–148.

Creasy, C.L. & Chernoff, J. (1995): Cloning and characterization of a human protein kinase with homology to Ste20. J Biol Chem **270**:21695–21700.

Darnell, J.E. (1978): Implications of RNA-RNA splicing in evolution of eukaryotic cells. Science **202**:1257–1260.

Doolittle, W.F. (1978): Genes in pieces: where they ever together? Nature **272**:581–582.

Gamulin, V., Skorokhod, A., Kavsan, V., Müller, I.M. & Müller, W.E.G. (1997): Experimental indication against blockwise evolution of metazoan protein molecules: example, receptor tyrosine kinase gene from the sponge *Geodia cydonium*. J Molec Evol **44**:242–252.

Geer, P., Hunter, T. & Lindberg, R.A. (1994): Receptor protein-tryosine kinases and their signal transduction pathways. Annu Rev Cell Biol **10**:251–337.

Gilbert, W. (1987): The exon theory of genes. Cold Spring Harbor Symp Quant Biol **52**:901–905.

Gilbert, W., Marchionni, M. & McKnight, G. (1986): On the antiquity of introns. Cell **46**:151–154.

Greenberg, M.E. & Ziff, E.B. (1984): Stimulation of 3T3 cells induces transcription of the c-*fos* proto-oncogene. Nature **311**:433–438.

Hanks, S.K. & Quinn, A.M. (1991): Protein kinase database: identification of conserved features of primary structure and classification of family members. Methods Enzymol **200**:38–62.

Hanks, S.K., Quinn, A.M.& Hunter, T. (1988): The protein kinase family: conserved features and deduced phylogeny of the catalytic domains. Science **241**:42–52.

Hardie, G. & Hanks, S. (1995): The Protein Kinase FactsBook: Protein-Tyrosine Kinases. Academic Press, London.

Hunter, T. (1987): A thousand and one protein kinases. Cell **50**:823–829.

Hunter, T., Lindberg, R.A., Middlemas, D.S., Tracy, S. & Geer, P. (1992): Receptor protein kinases and phosphatases. Cold Spring Harbor Symp Quant Biol **58**:25–41.

Koziol, C., Leys, S.P., Müller, I.M.& Müller, W.E.G. (1997): Cloning of Hsp70 genes from the marine sponges *Sycon raphanus* (Calcarea) and *Rhabdocalyptus dawsoni* (Hexactinellida). An approach to solve the phylogeny of sponges. Biol J Linnean Soc **62**: 581 – 592.

Kruse, M., Leys, S.P., Müller, I.M.& Müller, W.E.G. (1998): Phylogenetic position of hexactinellida within the phylum Porifera based on amino acid sequence of the protein kinase C from *Rhabdocalyptus dawsoni*. J. Molec Evol **46**: 721 – 728.

Kruse, M., Gamulin, V., Cetkovic, H., Pancer, Z., Müller, I.M. & Müller, W.E.G. (1996): Molecular evolution of the metazoan protein kinase C multigene family. J Molec Evol **43**:374–383.

Larson, J.S., Schuetz, T.J. & Kingston, R.E. (1988): Activation *in vitro* of sequence-specific DNA binding by a human regulatory factor. Nature **335**:372–375.

Lipps, J.H. & Signor, P.W. (eds) (1992): Origin and Early Evolution of Metazoa. Plenum Press, New York.

Loomis, W.F. (1988): Four Billion Years: An Assay on the Evolution of Genes and Organisms. Sinnauer Ass. Publishers, Sunderland Mass.

Mackie, G.O. & Singla, C.L. (1983): Studies on hexactinellid sponges. I, Histology of *Rhabdocalyptus dawsoni* (Lambe, 1873). Phil Trans R Soc Lond B **301**:365–400.

Mehl, D. & Reiswig, H.M. (1991): The presence of flagellar vanes in choanomeres of Porifera and their possible phylogenetic implications. Z zool Syst Evolut-Forsch **29**:312–319.

Morimoto, R. (1993): Cells in stress: transcriptional activation of heat shock genes. Science **259**:1409–1410.

Müller, W.E.G. (1995): Molecular phylogeny of metazoa [animals]: monophyletic origin. Naturwiss **82**:36–38.

Müller, W.E.G. (1997a): Molecular phylogeny of Eumetazoa: experimental evidence for monophyly of animals based on genes in sponges [Porifera]. Progr Molec Subcell Biol **19**:98–132.

Müller, W.E.G. (1997b): Origin of metazoan adhesion molecules and adhesion receptors as deduced from their cDNA analyses from the marine sponge *Geodia cydonium*. Cell & Tissue Res **289**: 383 – 395.

Müller, W.E.G. & Müller, I.M. (1997): Transition from Protozoa to Metazoa: an experimental approach. Progr Molec Subcell Biol **19**:1–22.

Müller, W.E.G., Müller, I.M. & Gamulin, V. (1994): On the monophyletic evolution of the Metazoa. Brazil J Med Biol Res **27**:2083–2096.

Müller, W.E.G., Kruse, M., Koziol, C. & Leys, S.P. (1998): Evolution of early Metazoa: Phylogenetic status of the Hexactinellida within the phylum of Porifera [sponges]. Progr Molec Subcell Biol **21**: 141 – 156.

Orgel, L.E. & Crick, F.H. (1980): Selfish DNA: the ultimate parasite. Nature **284**:604–607.

Ottilie, S., Raulf, F., Barnekow, A., Hannig, G. & Schartl, M. (1992): Multiple *src*-related kinase genes, *srk*1–4, in the fresh water sponge *Spongilla lacustris*. Oncogene **7**:1625–1630.

Palmer, R.H., Ridden, J. & Parker, P.J. (1995): Cloning and expression patterns of two members of a novel protein-kinase-C-related kinase family. Eur J Biochem **227**:344–351.

Patthy, L. (1994): Introns and exons. Curr Opin Struct Biol **4**:383–392.

Patthy, L. (1995): Protein Evolution by Exon-Shuffling. Springer Verlag, New York.

Reiswig, H.M. & Mackie, G.O. (1983): Studies on hexactinellid sponges. III The taxonomic status of the Hexactinellida within the Porifera. Phil Trans R Soc Lond B **301**:419–428.

Rousset, D., Agnès, F., Lachaume, P., André, C. & Galibert, F. (1995): Molecular evolution of the genes encoding receptor tyrosine kinase with immunoglobulinlike domains. J Mol Evol **41**:421–429.

Schäcke, H., Müller, W.E.G., Gamulin, V. & Rinkevich, B. (1994a): The Ig superfamily includes members from the lowest invertebrates to the highest vertebrates. Immunology Today **15**:497–498.

Schäcke, H., Rinkevich, B., Gamulin, V., Müller, I.M. & Müller, W.E.G. (1994b): Immunoglobulinlike domain is present in the extracellular part of the receptor tyrosine kinase from the marine sponge *Geodia cydonium*. J Molec Recognition **7**:272–276.

Schäcke, H., Schröder, H.C., Gamulin, V., Rinkevich, B., Müller, I.M. & Müller, W.E.G. (1994c): Molecular cloning of a receptor tyrosine kinase from the marine sponge *Geodia cydonium*: a new member of the receptor tyrosine kinase class II family in invertebrates. Molec Membrane Biol **11**:101–107.

Scheffer, U., Krasko, A., Pancer, Z. & Müller, W.E.G. (1997): Isolation and characterization of the cDNA clone encoding the serum response factor homolog in the marine sponge *Geodia cydo-*

nium: high conservation of this transcription factor within Metazoa. Biol J Linnean Soc, in press.

Seimiya, M., Naito, M., Watanabe, Y. & Kurosawa, Y. (1997): Homeobox genes in the freshwater sponge *Ephydatia fluviatilis.* Progr Molec Subcell Biol **19**:133–155.

Seino, S., Seino, M., Nishi, S. & Bell, G.I. (1989): Structure of the human insulin receptor gene and characterization of its promoter. Proc Natl Acad Sci USA **86**:114–118.

Shore, P. & Sharrocks, A.D. (1995): The MADS-box family of transcription factors. Europ J Biochem **229**:1–13.

Solé-Cava, A.M. & Thorpe, J.P. (1991): High levels of genetic variation in natural populations of marine lower invertebates. Biol J Linn Soc **44**:65–80.

Stolzfus, A. (1994): Origin of introns – early or late? Nature 369:526–527.

Sun, H. & Tonks, N.K. (1994): The coordinated action of protein tyrosine phosphatases and kinases in cell signaling. Trends Biochem Sci **19**:480–485.

Ullrich, A. & Schlessinger, J. (1990): Signal transduction by receptors with tyrosine kinase activity. Cell **61**:203–212.

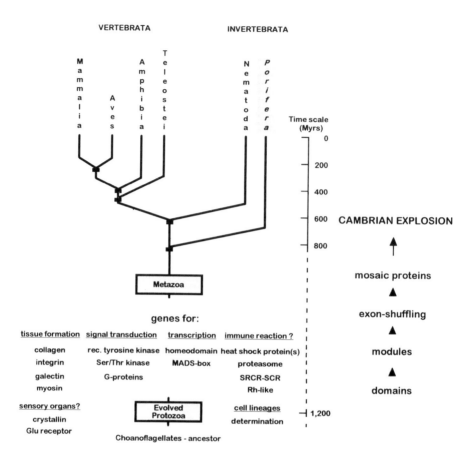

Fig. 1. Phylogenetic relationship of Porifera within the animal groups based on molecular biological data obtained from sequences of "metazoan" proteins required for tissue formation, signal transduction, transcription, immune reaction [potential] and sensory organ [potential]. The cell lineages in sponges are less determined than in higher Metazoa. It is proposed that the Cambrian explosion of metazoan radiation became possible after the creation of the evolutionary mechanism of modularization of distinct protein domains, thus allowing the formation of mosaic proteins by exon-shuffling; this process happened approximately 1,000 MYA. It is assumed that Metazoa originated from evolved Protozoa, e.g. Choanoflagellata.

A   RT-Kinase_gene

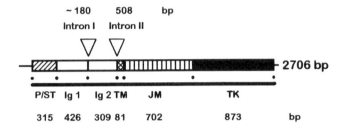

B   nS/T-Kinase_gene

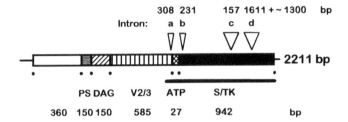

C   cS/T-Kinase_gene

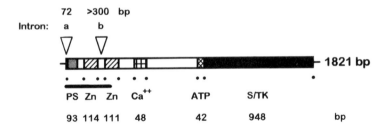

Fig. 2. Structure of the receptor Tyr-kinase and Ser/Thr-kinases from *G. cydonium*. **A.** RT-Kinase_gene: The building blocks of the deduced aa sequence of the RTK are shown: Pro-Ser-Thr[P/S/T]-rich domain, Ig-like domains 1 and 2 [Ig 1 and Ig 2], transmembrane domain [TM], the juxtam-embrane region [JM] and the TK-domain [TK]. **B.** nS/T-Kinase_gene [partial]. The deduced aa se-quence of the "novel" [Ca^{2+}-independent] protein kinase C [nPKC] displays the following boxes characteristic for nPKC; the Ser/Thr-kinase catalytic domain [S/TK], the putative ATP-binding do-main [ATP], the pseudosubstrate segment [PS] and the phorbol esters/diacylglycerol binding do-main [DAG]. **C.** cS/T-Kinase_gene [partial]. The following units are present; the pseudosubstrate segment [PS], two zinc fingers [Zn], the typical phorbol esters/diacylglycerol binding domain, the C2 domain signature, which is likely involved in Ca^{2+} binding [Ca^{2+}], an ATP/GTP-binding site motif [ATP] and the Ser/Thr-kinase catalytic domain [S/TK]. The numbers below the schemes of the respective genes indicate the sizes of the respective coding segments in bp; the numbers at the end refer to the sizes of the total coding regions. The positions of the introns and their sizes are given. The regions within the genes which have been sequenced are delimited [——].

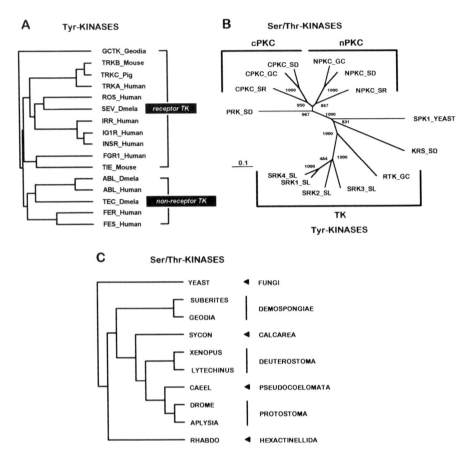

Fig. 3. A. Dendrogram deduced from multiple alignment of the TK-domain of RTK gene of *G. cydonium* [*GCTK*_Geodia] with those of ten receptor TKs [neurotrophin-4 receptor [*TRKB*_Mouse]; NT-3 growth factor precursor [*TRKC*_Pig]; nerve growth factor receptor [*TRKA*_Human]; ROS proto-oncogene tyosine kinase [*ROS*_Human]; *D. melanogaster* sevenless receptor PTK [*SEV*_Dmela]; insulin receptor-related receptor [*IRR*_Human]; insulin-like growth factor 1 receptor precursor [*IG1R*_Human]; insulin receptor precursor [*INSR*_Human]; basic fibroblast growth factor receptor 1 precursor [*FGR1*_Human] and TIE protein-Tyr-kinase [*TIE*_Mouse]] and of five non-receptor TKs [*DASH/ABL* proto-oncogene Tyr-kinase from *D. melanogaster* [*ABL*_Dmela]; ABL proto-oncogene Tyr-kinase [*ABL*_Human]; *D. melanogaster* SRC protein Tyr-kinase [*TEC*_Dmela]; FES/FPS-related PTK [*FER*_Human]; FES/FPS protein-Tyr-kinase [*FES*_Human]]. **B.** Phylogenetic tree [unrooted] of the catalytic domains built from following kinases; *cPKC* from *G. cydonium* [CPKC_GC], *S. domuncula* [CPKC_SD] and *S. raphanus* [CPKC_SR], *nPKC* from the same animals NPKC_GC, NPKC_SD and NPKC_SR; *KRS* from *S. domuncula* [KRS_SD], *PKR* from this sponge species [PRK_SD] and the Tyr-kinase, the *RTK* from *G. cydonium* [RTK_GC]. In addition, the *scr*-related kinases from the sponge *Spongilla lacustris* [SRK1_SL, SRK2_SL, SRK3_SL, SRK4_SL] and the Ser/Thr and Tyr phosphorylating kinase SPK1 from *Saccharomyces cerevisia* [SPK1_SC] have been included. Calculation was performed by neighbour-joining. The numbers at the nodes refer to the respective level of confidence [1000 bootstrap replicates]. Scale bar indicates the evolutionary distances. **C.** Unrooted phylogenetic tree computed from ten Ser/Thr-kinase sequences [catalytic domain; cPKCs]. The following sequences have been used: *I. Metazoa*, cPKC from the *deuterostomes Xenopus laevis* [XENOPUS] and *Lytechinus pictus* [LYTECHINUS], from the *protostomes* cPKC from *D. melanogaster* [DROME] and *Aplysia californica* [APLYSIA] and those from the *sponges* of the classes *(i)* Demospongiae, *G. cydonium* [GEODIA] and *S. domuncula* [SUBERITES], *(ii)* Calcarea, *S. raphanus* [SYCON], and *(iii)* the Hexactinellida, *Rhabdocalyptus dawsoni* [RHABDO] as well as from *II. Fungi Saccharomyces cerevisiae* [YEAST].

Receptor tyrosine kinase - serum response factor

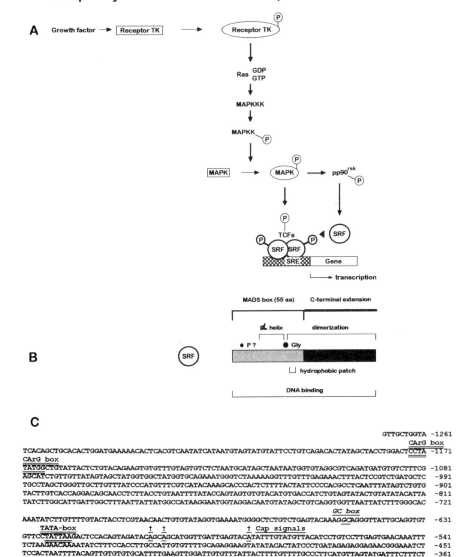

C

```
                                                                  GTTGCTGGTA -1261
                                                                     CArG box
TCACAGCTGCACACTGGATGAAAAACACTCACGTCAATATCATAATGTAGTATGTATTCCTGTCAGACACTATAGCTACCTGGACTCCTA -1171
CArG box
TATGGCTGTATTACTCTGTACAGAAGTGTGTTTGTAGTGTCTCTAATGCATAGCTAATAATGGTGTAGGCGTCAGATGATGTGTCTTTCG -1081
AGCATCTGTTGTTATAGTAGCTATGGTGGCTATGGTGCAGAAATGGGTCTAAAAAGGTTTGTTTGAGAAACTTTACTCCGTCTGATGCTC  -991
TGCCTAGCTGGGTTGCTTGTTTATCCCATGTTTCGTCATACAAAGCACCCACTCTTTTACTATTCCCCACGCCTCAATTTATAGTCTGTG  -901
TACTTGTCACCAGGACAGCAACCTCTTACCTGTAATTTTATACCAGTAGTGTGTACATGTGACCATCTGTAGTATACTGTATATACATTA  -811
TATCTTGGCATTGATTGGCTTTAATTATTATGGCCATAAGGAATGGTAGGACAATGTATAGCTGTCAGGTGGTTAATTATCTTTGGGCAC  -721
                                                                  GC box
AAATATCTTGTTTTGTACTACCTCGTAACAACTGTGTATAGGTGAAAATGGGGCTCTGTCTGAGTACAAAGGCAGGGTTATTGCAGGTGT  -631
    TATA-box                          † †                    † Cap signals
GTTCCTATTAAGACTCCACAGTAGATACAGCAGCATGGTTGATTGAGTACATATTTGTATGTTACATCCTGTCCTTGAGTGAACAAATTT  -541
TCTAAGAACAAAATATCTTTCCACCTTGCCATTGTGTTTTGCAGAGGAAGTATATACACTATCCCTGATAGAGAGGAGAACGGGAAATCT  -451
TCCACTAATTTTACAGTTGTGTGTGCATTTTGAAGTTGGATTGTGTTTATTACTTTTGTTTTGCCCTTCATGTTAGTATGATTTCTTTCT  -361
CTTGTTCAACAGTTTGGACACAGCTCTGTCTTCATTGTCTTCCTTCGTGTGACTGTAGTGACTGTCTCCTCTCAAGATTGCCAGAGAGGT  -271
AAAGAAATTGGGCTATCGTGGGTTTCTTTTTAGAGCATGGTTTTCTTCTTTCTCAAGTTCTTTCAGGGCTTCAAGATGTTGGACCTACAG  -181
TACCTCCAGTGACATACAATGGAATTGTGCCAAGAATATCACCATGTGGAGCCTCTCCAACAAGTTTGAATTGCAAGGCAGGAGAAGGAG   -91
TGTTGAGTGATGGGAACATTGGACCTACAGACTCAGTGGATGTCAGTGATCCAGACCAGGTCACCAGGTTCTTTGTCTGGCATAGAGACA    -1
```

Fig. 4. A. Postulated signal transduction pathway in *G. cydonium* with the receptor Tyr-kinase [Receptor TK; RTK] as signal receiver and the serum response factor [SRF] as effector. **B.** Scheme of the DNA-binding domain of the SRF with its conserved MADS-box motif and the C-terminal extension which promotes dimerization. Further details about this signalling cascade are given in the text. **C.** Nucleotide sequence of the putative promotor of the RTK gene. The potential transcription starts at nt +1; from there the promotor nts are number with a negative prefix. The GC box [italics and underlined], the TATA-box [double underlined] the cap signals [† and underlined] and the CArG box-like sequence [double underlined] are marked.

```
                                                               GTTGCTGGTA -1261
TCACAGCTGCACACTGGATGAAAAACACTCACGTCAATATCATAATGTAGTATGTATTCCTGTCAGACACTATAGCTACCTGGACTCCTA -1171
        DfD>(89.6%)
TATGGCTGTATTACTCTGTACAGAAGTGTGTTTGTAGTGTCTCTAATGCATAGCTAATAATGGTGTAGGCGTCAGATGATGTGTCTTTCG -1081

AGCATCTGTTGTTATAGTAGCTATGGTGGCTATGGTGCAGAAATGGGTCTAAAAAGGTTTGTTTGAGAAACTTTACTCCGTCTGATGCTC  -991

TGCCTAGCTGGGTTGCTTGTTTATCCCATGTTTCGTCATACAAAGCACCCACTCTTTTACTATTCCCCACGCCTCAATTTATAGTCTGTG  -901
                                                                   CF2-II>(90%)
TACTTGTCACCAGGACAGCAACCTCTTACCTGTAATTTTATACCAGTAGTGTGTACATGTGACCATCTGTAGTATACTGTATATACATTA  -811
                DfD>(87.1%)                                           DfD>(85.1%)
TATCTTGGCATTGATTGGCTTTAATTATTATGGCCATAAGGAATGGTAGGACAATGTATAGCTGTCAGGTGGTTAATTATCTTTGGGCAC  -721

AAAATATCTTGTTTTGTACTACCTCGTAACAACTGTGTATAGGTGAAAATGGGGCTCTGTCTGAGTACAAAGGCAGGGTTATTGCAGGTGT  -631
                                                                      <HSF(100%)
GTTCCTATTAAGACTCCACAGTAGATACAGCAGCATGGTTGATTGAGTACATATTTGTATGTTACATCCTGTCCTTGAGTGAACAAATTT  -541
    HSF>(88.5%)<HSF(85.9%)                     CF2-II>(90%)           HSF>(88.5%)   HSF>(85.9%)
TCTAAGAACAAAATATCTTTCCACCTTGCCATTGTGTTTTGCAGAGGAAGTATATACACTATCCCTGATAGAGGAGAACGGGAAATCT  -451
                                 DfD>(90.6%)                     <HSF(100%)<HSF(100%)
TCCACTAATTTTACAGTTGTGTGTGCATTTTGAAGTTGGATTGTGTTTATTACTTTTGTTTTGCCCTTCATGTTAGTATGATTCTTTCT  -361

CTTGTTCAACAGTTTGGACACAGCTCTGTCTTCATTGTCTTCCTTCGTGTGACTGTAGTGACTGTCTCCTCTCAAGATTGCCAGAGAGGT  -271
HSF>(100%)          <HSF(100%)            <HSF(100%)<HSF(100%)<HSF(95.3%)<HSF(88.5%)
AAAGAAATTGGGCTATCGTGGGTTTCTTTTTAGAGCATGGTTTTCTTCTTTTCTCAAGTTCTTTCAGGGCTTCAAGATGTTGGACCTACAG -181
TACCTCCAGTGACATACAATGGAATTGTGCCAAGAATATCACCATGTGGAGCCTCTCCAACAAGTTTGAATTGCAAGGCAGGAGAAGGAG  -91
TGTTGAGTGATGGGAACATTGGACCTACAGACTCAGTGGATGTCAGTGATCCAGACCAGGTCACCAGGTTCTTTGTCTGGCATAGAGACA   -1
```

Fig. 5. Identification of elements in the promotor of the RTK for regulatory transcription factors, e.g. for binding of the heat shock factors [HSFs], heat shock elements [HSEs; marked underlined], for homeodomains [here: DfD; in italics and double underlined] and for the CF2-II factor [double underlined]. The direction of binding is indicated [< or >].

```
GCnPK1ge AGCTTCCACGGCTTTGGGCGACCGGCGATGAAGAAGTACAAACTGGAGCAGTTCAAGTTTCTCAAACTCCTTGGGAAAGGAAGCTTTGGG  1260
         SerPheHisGlyPheGlyArgProAlaMetLysLysTyrLysLeuGluGlnPheLysPheLeuLysLeuLeuGlyLysGlySerPheGly      420
GCnPK1ge AAGGTAATGGAGACACATATTATGTTTAGTGAGTGAATATGAGAATTCCCAATGATGGTACAATGTCTTAGTTTAAGTATACACATCGCA  1350
         Lys                                                                                           421
GCnPK1ge ACTTTAGTGAGAGGTAGCAAAATCTTCAGCATTTTGGCAAAGAAAGGGCTGGTGCAACTTCAATTGTATATGTGTGATGGGACTTTTTGC  1440
GCnPK1ge ATGAGTGCATGGGTAAATATGAGCGCGTAACTTTTTCACGGTAAGCATGTTACTTTTGACAGTTTGTGTTATGTGTTTAAGTTGTGCTGT  1530
GCnPK1ge GGCATCGTGATCTCTTTCTGCGCTGTACATGTATGTACATAGGTCTTGCTGGCCCAACTGGAGGGGGAATGAGCAGTATTTTGCCATCAAG  1620
                                    ValLeuLeuAlaGlnLeuGluGlyAsnGluGlnIleTyrPheAlaIleLys                 437
GCnPK1ge GCCCTCAAGAAGGACGTAGTACTGGAGGACGATGACGTGGAGGCCACTATGGTGGAGAAGAGACTTCTCGCTCTTGGCTGTAACCATCCT  1710
         AlaLeuLysLysAspValValLeuGluAspAspAspValGluAlaThrMetValGluLysArgLeuLeuAlaLeuGlyCysAsnHisPro     467
GCnPK1ge TTCCTCACTCACCTCCACTCCACCTTCCAGACCCCCGTGAGTCCGTCTTTTCCTTCCATTCTTTCACCTTTTTCTTCATTCAACATACCT  1800
         PheLeuThrHisLeuHisSerThrPheGlnThrPro                                                          479
GCnPK1ge TTTAGACACCCTCTGAGCCTCTCTCTATCTTCTTTCCTTCCATCCTGTCCATCCCCTTACACTCCCCTGTCAATTCAGCGAAGCACTTTT  1890
GCnPK1ge ACAATAGAGTAAAACTCCTTTCTCGCACCTTCCCTTCCTCTTTTTCTGCATCACAGAATTCAACCCTCTTTCCCCTCTCTGTTTCAGAGT  1980
                                                                                                Ser   480
GCnPK1ge CACCTGTTCTTTGTGATGGAGTATCTGAATGGCGGCGATCTCATGTATCACATACAGATTTCTCACAAATTCAAACTCCCCAGAGCAAGG  2070
         HisLeuPhePheValMetGluTyrLeuAsnGlyGlyAspLeuMetTyrHisIleGlnIleSerHisLysPheLysLeuProArgAlaArg     510
GCnPK1ge TTCCATGCTGCAGAAATACTCTGTGCTCTTCAGTTCCTCCACAAACAAGGCATTATATACAGAGATCTCAAGTTGGACAACGTGATACTG  2160
         PheHisAlaAlaGluIleLeuCysAlaLeuGlnPheLeuHisLysGlnGlyIleIleTyrArgAspLeuLysLeuAspAsnValIleLeu     540
GCnPK1ge GACTCTGAGGGTCACTGTAAACTGGCCGACTTTGGCATGTGCAAGGAGAACATCATTGGGTATGCCACTGCAGGCACCTTCTGTGGGACA  2250
         AspSerGluGlyHisCysLysLeuAlaAspPheGlyMetCysLysGluAsnIleIleGlyTyrAlaThrAlaGlyThrPheCysGlyThr     570
GCnPK1ge CCAGACTACATATCACCAGAGATCATAAAGGGGAAGAGGTATACATTCTCTGTGGACTGGTGGTCTTTTGGAGTCCTCTGCTACGAGATG  2340
         ProAspTyrIleSerProGluIleIleLysGlyLysArgTyrThrPheSerValAspTrpTrpSerPheGlyValLeuCysTyrGluMet     600
GCnPK1ge ATTACCGGCCAGGTGAGTGGTTAGACAATGTTATGCAACAGCATGCTTTTAACTCCTAAAGAACTTTAATACCATTCAATCCTTCAAAAC  2430
         IleThrGlyGln                                                                                  604
GCnPK1ge CTCATGCTCCGTCAATAGATGTATGTGGTAGTTAAGACCAAGTATGTCAGTGTATGAGTGTGTTGTGTCTGTGTGCTGCAGTCTCCATTCAG  2520
                                                                        SerProPheSe                    608
GCnPK1ge TGGAGAGGATGAGGACGAGTTGTTTGACTCAATCTGCAACCATCAGGTCTCCTTCTCTCGCTACCTCGACCAAACCACCCATCAACTTCCT  2610
         rGlyGluAspGluAspGluLeuPheAspSerIleCysAsnHisGlnValSerPheSerArgTyrLeuAspGlnThrThrIleAsnPheLe      638
GCnPK1ge TGACAAGGTATATAATAGTTAGGTACTTTGACGAATACGTATTTAGGCTGAGAGTAAAGTGCAGTCACTCTACAAAAGTCTTATTTAGGA  2700
         uAspLys                                                                                       640
GCnPK1ge CTACTATGCACCCACTCACTTGTGTTCATCTATAACAAGAAGCCCAAGCCCTTTCCTTTTAATATAGTCCACCTACGTTATCCCCTAAA  2790
GCnPK1ge CAGGGACATAGTAATCTATAAGGTTATTCAGAATGAGTGCATGAGTTGTAAGATAGATCGAGCTGCCGTTCAGCCAGGCGCTGGAGATGA  2880
GCnPK1ge TCAAAACCGGCGAGATACGTGACGGTAAGACGGTGTTATTGCTTAACTATTTGCAAACGTCACATTTAATGAGCTGAAAAATAACAATAA  2970
GCnPK1ge TATTTCGTTGTTTATTATTGGCTCAATCCGACAAATCCGATTGAGCCGCGACTCGTCTGCGCAACGAAGATACGACTTGTGCTGTTTGTTTG  3060
GCnPK1ge AACTTCTGGGGTCGTACCGTCCATGCGCTATCGCATTTCCTTCTCTTTTTTTTCGCTTTGTTGCCGACGTCTTTGGTGTGGGCGGCACC  3150
GCnPK1ge AGCGCAACGGGCGTTTTCCGACTGGCAGGTCACCTGTAATAACCAAAATTTCTGCGTGGCGGTAATACGGGCGATCATAATGGACTGGT  3240
GCnPK1ge GATGACCCTGAGCCGCAGCGCCGGGGCGCATACCGATGCCGTTTTACGTATTGAGCGCGGCGGATTGAAGTCGCCGGAGGCGTCAGAAGG  3330
GCnPK1ge GGAGATAGCGCCACGGCTGCTGTTAGATGGCGAGCCGTTGGCGTTAAGTAGCTGACAAGTGGCGGATTTCACCATGGTTATTAGTAACCGA  3420
GCnPK1ge TGATACGGCAACCATCACCGCGTTTTTGCAGATGATTCAGGAAGGGAAGGCAATCACCTTACGCGATGGCGATCAGACCATTTCTCTGAG  3510
GCnPK1ge TGGCTTAAAAGCAGCGTTGTTGTTTATTGATGCTCAGCAAAAGCGCGTTGGCAGTGAAACCGCGTGGATCAAGAAAGGGGACGAACCGCC  3600
GCnPK1ge GCTCAGCGTACCGCCCGCGGCCTGCGCTGAAAGAGGTCGCGGTGGTTAACCCAACGCCGACCACTCTCACTCGAAGAACGCAACGATTT  3690
GCnPK1ge GCTGGATTATGGAAACTGGCGGATGAATGGTCTGCGCTGCTCGCTTGATCCATTGCGTCGTGAGGTGAATGTCACTGCGCTGACTGATGA  3780
GCnPK1ge TAAAGCGCTGATGATGATTAGCTGTGAGGCAGGGCCTATAACACCATTGATTTGGCATGGATTGTGTCGCGTAAAAGCCACTACGTTC  3870
GCnPK1ge GCGCCCGGTTCGGTTGCGTTTGCCGTTCAACAACGGTCAGGAGACGAATGAACTGGAACTGATGAAGCGCAACATTTGATGAGAAATCGCG  3970
GCnPK1ge TGAACTGGTGACCTTAGCGAAAGGGCGCGGATTAAGCGATTGTTGGCATTCAGGCGCGCTGGCGCTTTGACGGTCAGCGTTTTCGCCTGGT  4050
GCnPK1ge GCGTTACGCCGCAGAACCCACCTGCGATAACTGGCATGGGCCAGATGACGCTTGGCCACGTTGTGGATCACCCGTTAAACAAAATGGCGGGAC  4140
GCnPK1ge ATGTAGGCCGGATAAGACGCTGTAGGCCGGATAAGACGCGTCAGCGTCGCATCCGGCAATCAGCATCCGGCAATCACCATCAGGCACC--  4228
GCnPK1ge 1.3 kb --TTGTTGCAGAGAGATCCAGGGGAGAGGCTGGGCTTGCATCGAAGACAGAGAGAATATCAGAGCACACGCCTTCTTCAGAG  4318
         LeuLeuGlnArgAspProGlyGluArgLeuGlyLeuHisArgArgGlnArgGluTyrGlnSerThrArgLeuLeuGlnAr              667
GCnPK1ge AAATTTGACTGTTGTCAAACTGGAGGCCAGGAAACTAAAACCTCCCTTCAAACCCAATGTGAAAGGGGCGTCAGATGCCAGCAACTTTGG  4408
         gAsnLeuThrValValLysLeuGluAlaArgLysLeuLysProProPheLysProAsnValLysGlyAlaSerAspAlaSerAsnPheGl     697
GCnPK1ge CGATGATTTCACCTTCCAGCCAGCCCAGCTCACTCCCACTGACACCACACTGGTGATGTCCATCGACCAAACAAACTTCACCGGGTTCTC  4498
         yAspAspPheThrPheGlnProAlaGlnLeuThrProThrAspThrThrLeuValMetSerIleAspGlnThrAsnPheThrGlyPheSe     727
GCnPK1ge ATTCACTAGTGACCTCTATTCCAAACTCCACTGATATTATCTAACRCTGCTATATATTATAATTGTAATGATTGTGATCTGCCTAAGGCT  4588
         rPheThrSerArgLeuTyrSerLysLeuHis                                                               737
GCnPK1ge CAGCTGTAAGTAGAGCGTCTCTCATTTTCTCTGTTTTGTAAGTGCTGCTACAGTATGCTGACAATGTGTTTTTAACTGATGTGATCATTT  4678
GCnPK1ge TTTATCCCGAGTGAGTGTATTGTACAATAATAAATGTTGTGTTA$_n$                                               4732
```

Fig. 6. Sequence of nS/T-Kinase_gene covering the Ser/Thr-kinase domain. The nt sequence numbering starts with those identified in the cDNA (Kruse et al. 1996); subsequently also the nts of the introns are included. The coding regions with the deduced aa are shown; the intron nts are marked in italics and are underlined.

Werner E.G. Müller e.a.

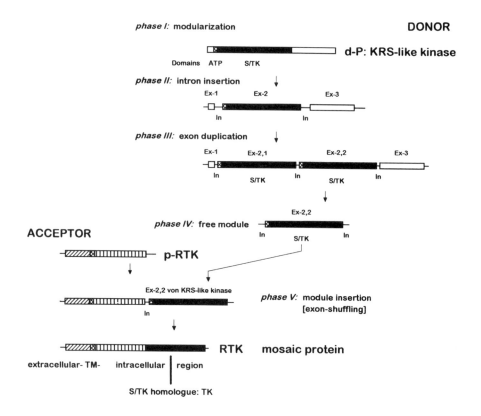

Fig. 7. Proposed generation of the mosaic protein RTK. The gene coding for a ***donor protein*** d-P – here the KRS-related kinase – consisting of an ATP binding domain [ATP] and the Ser/Thr-kinase domain [S/TK] [phase I] undergoes modularization [phase II] by insertion of introns [In]. One of the created exons, Ex-2 is duplicated [phase III] and its poduct, Ex-2,2 is liberated [phase IV] and inserted as a module into the gene of an ***acceptor protein***, here the precursor gene of RTK [p-RTK]; the process of module insertion by exon-shuffling is completed [phase V] and the mosaic protein – RTK – formed (modified after Patthy 1994 and Müller and Müller 1997).

The evolution of the mollusc construction:
Living organisms as energy-transforming systems

Karl Edlinger

Introduction

Modern science, especially physics and chemistry as the most prominent disciplines, are based on the elucidation of principles of high abstractness. Corroboration of explanations is usually obtained by experiments and structural analysis.

For a long time biology was an exception to these rules. Biologists were engaged in the business of describing living beings without giving a theoretical foundation. As a consequence the very prominent subject of biology, the living organism, was unknown as a theoretically conceived entity.

In previous years biological theories were presented, which meet the requirements of theoretically well founded natural sciences. As a result of theoretically-based constructional analyses, living organization appears as the expression of law-like principles and ruled by natural laws. As a consequence, organisms can only be understood as wholes and as permanently active systems, kept alive by incessant energy transformation (Gutmann 1988, 1989, Gutmann & Edlinger 1991a, b; 1994a, b, c, d; Edlinger, Gutmann & Weingarten 1989).

The permanent activity and energy transformation presuppose a construction which is sophisticated in a high degree. All parts at all levels of the organismic constructions must fit to each other and to the energy converting constructions as wholes (Edlinger 1991c, d). Constructional constraints are based on the properties and on the arrangements of the parts of the construction, which interact mutually. This interaction concerns physical mechanisms as well as chemical ones. So the construction can determine all details on all levels of living beings and is influenced and determined by these for its own (Fig. 1). Constraints are also determining, whether some details of structures of organisms can persist during evolution or whether they can change their functions, or must disappear. Even the features at the submicroscopical level are constrained by the constructional requirements and functions of the living machine and by the demands of energy transformation (Bereiter-Hahn 1991; Edlinger 1994b).

From this, it follows that evolutionary trends on the histological and ultrastructural levels are partly determined by the constitution and the evolution of the constructional frame in which all partial systems must be delicately integrated.

Hence alterations of numerous morphological or ultrastructural features cannot be conceived as isolated processes. Their special trends depend always on constructional laws of the organism as a whole and also of its subservient functional parts. Because of very strict interconnections and interdependencies of most partial systems alterations of some structural details must necessarily bring about alterations of other parts. As a consequence, the application of terms like plesiomorphy or apomorphy in systematics causes misunderstanding instead of giving reliable reasons for the reconstruction of evolutionary transformations.

On the basis of the outlined presuppositions and implicit theorems the misunderstanding of traditional morphology even of submicroscopical morphology, which did not expect the determining effect of physical and mechanical laws on living organization, is suspended. In the light of organismic explanation even the hazards of mutations are considerably restrained by constructional limitations.

As principles for constructional transformation can be given, evolution is no longer conceivable as a tinkering process or as a mosaic alteration. New consistent models for the explanation of bauplan organization and organizational transformation in evolution can be reconstructed. Traditional "gestalt"-sequences are replaced by model-like explanatory reconstructions.

Evolutionary changes in the transition to the mollusc construction

Mollusks are creeping creatures. In the primitive constructions the creeping foot follows the form of the substrate and, as a consequence, develops adhesion to the substrate. Such a kind of motility is neither primitive nor accessible for all soft body constructions as it presupposes a dense tethering of muscles in the foot. For the incipient stages of the transition to the mollusc construction, which must have consisted of worm-like organisms, a capability of flattening the ventral body side was an indispensable prerequisite (Gutmann 1974, Edlinger 1991a). This capability of flattening the ventral and concomitantly the dorsal side is guaranteed only in metamerically tethered forms and effected by the hydraulic apparatus. Muscle bearing dissepiments and muscles which traverse the inner hydraulic filling allow variation of the cross-section as a consequence of growing density of the tethering muscles and of the adhesion to the substratum (Fig. 2).

While the ventral portion of the body wall developed a highly deformable muscle lattice the dorsally shifted parts of these constructions persisted as metameric coelomate constructions with a tube-like intestinal canal, segmented muscles, nephridia, and coelomic chambers. The coelom continued to surround the gut and enforced its tube-like form. It also enclosed the heart with a portion that functioned as a pericardium. This specialized portion of the coelom enabled the gliding of the permanently beating heart.

In the thickened ventral wall of the body metamerism disappeared. A lattice-like tethering system, consisting of musculature and connective tissue, was established as a highly deformable organ. The capability of flattening of the so called foot was the result of the step by step transformations. The foot was able to actively adapt to the surface-structures of hard substrate. In conjunction with the muscle system of the foot mucus glands on the sole were formed. The mucus, together with the contour following adaptation of the surface of the foot, ensured adhesion to the substrate.

Adhesion to the substrate enabled the mollusc-predecessors to use chitineous teeth around the mouth to function as a radula. When the animals were better fixed to the substrate by the flattened foot, the body could be used as a stabilizer of the radula apparatus in rasping activity at the substrate.

The interconnectedness of the functional subsystems in the formation of the mollusc body construction is apparent. Development of the foot as an isolated organ with very sluggish and slow movement would not have made sense without the relation to the radula. The radula alone would not have been feasible because scraping-attempts could not have been successfully performed because the body would have been pushed away from the substrate. The formation of the shell plates which must have also occurred during the early stages of mollusc evolution was an economizing alteration which was only possible in constructions which restrained peristaltic movements to the ventral side of the body. These evolutionary changes resulted in the suppression of motility and deformability of the dorsal parts of the body. In the undeformed portion the formation of dorsal shell structures in a meta-merically subdivided form was possible. The dorsal surface could be covered by a plate-formed secretion. The array of the elements of the later shell retained its sub-division in accordance with the structure of the metameric coelom.

From the incipient stages of development the shell plates could serve as an in-sertion for the dorsoventral musculature and exert a stabilizing influence on the dorsally located internal organs. This effect is part of the economizing effect of shell formation. The shell could replace energetically more expensive structures such as muscles and connective tissue tethering. While the retractor muscles persisted as serially arranged entities the muscles in the ventral body wall formed the unseg-mented tethering lattice of the foot. In the first stages of economizing shell forma-tion freedom for worm-like movements must have been guaranteed. So even poly-placophoran constructions retain a little bit of the deformability of the worm-like predecessors.

The lateral groove between the shell and the foot could accommodate the gills and the nephropori were arranged in a serial order. These organs were necessary because the dorsal and ventral surfaces were covered and no longer useable for respiration. Pari passu with the transformation of the body machine the coelomic system continued to lose its metameric organization. The intestinal canal would no longer have been stabilized in the vault of the shells when the coelom lost its hy-draulic function. The gut was now held in position by the gradually enlarging mid-gut glands which served as a cushion and replaced the coelomic chambers and the mesenteria as suspending structures.

The just outlined model gives a sketch for the evolutionary alteration of annelid-like ancestors into basic mollusk constructions. Mollusk evolution is based on two key transformation steps: Flattening of the worm-body and formation of foot thereby allowing adhesive creeping on one hand and enhancement of use of radula for scraping food from rocky substrate on the other hand. Both aspects are construc-tional and functionally interdependent.

The early chiton-like mollusks described above gave rise for two phylogentic branches visible in the recent mollusks too: The Amphineura, represented by the

Placophora and Solenogastres, and the conchiferans, which arose from high chambered chiton-like ancestors by fusion of the dorsal shell plates and the constructional changes which happened as a consequence of this fusion.

The solenogastres

From the early chiton-like stage of flexible elongate mollusk constructions the placophorans and the solenogastres must be derived. The coelomatic metamerism of the placophorans disappears by further reduction of the coelom and by an enlargement of the midgut-gland. But at the other side metamery persists in the musculature and in the subdivided shell. The shell provides muscle scars and so a serial arrangement of shell plates and dorsoventral musculature can be done mutually. At the other side blood vessels change their place, they are outside the dorsoventral musculature and as a consequence of the constructional needs metameric arrangement disappears. These placophoran constructions could be flattened in a high degree.

The Solenogastres also can be derived from coelomatic chiton-like constructions (Gutmann 1974, Edlinger 1991a) (Fig. 3). Their shell was reduced step by step and the girdle enlarged. In the case of the Caudofoveata the muscular foot was reduced to a digging apparatus in the rostral part of the construction. Only in this part the dorsoventral musculature persisted in its typical structure and arrangement. The gut is also enlarged. Because of the arrangement of the dorsoventral musculature gut diverticles arise. In the rear of the body the girdle enlarged and finally was closed ventrally.

The rear part was filled by an enlarged midgut-gland, surrounding a tub-like gut. Only in the posterior part a rest of the lateral grove, so to say of the mantle cavity, persisted with one pair of gills.

In the Ventroplicida the shell also was reduced and the girdle enlarged. But, in contrast to the caudofoveata, the foot of the Ventroplicida, persisted over the whole length of the animals. As a consequence the musculature which tethers the cross-section persists in these constructions. By reduction of the coelom a fluid-filled space arose, only partially filled by glandular structures of the gut. The rest of the inner space was filled by enlarged pouches of the gut.

The conchiferans

The conchiferans have to be considered as the result of a major transformation step which started with the fusion of the shell elements into an unique entity (Gutmann 1974, Edlinger 1989a, 1991a) (Fig. 2). The monoplacophorans demarcate the transition field to the conchiferan level of organization. Thickening of the foot was prerequisite for this stage. While the shell elements could fuse metamerism of the coelom, nephridia and gills is still evident. In connection with the fusion of the shell deepening of the waist in the soft body must have developed. Thereby the cephalopodium obtained a high degree of free motility while the inner organs were held undeformed far off the substrate (Edlinger 1988a, b, 1989a).

As a result of the transformation steps leading to the Conchifera, the inner organs had to undergo profound reconstructions. A reduction of space for the insertion of gills and the number of retractors enforced the gradual loss of the remaining coelomic structures and of metamerism. These structures are observable only in some recent neopilinida, excluding dwarfish forms, which had to undergo many changes because of some needs of mechanical construction and metabolism.

Transition stages with a clearly discernible radiation in the paleozoic sediments are well represented by fossils. All conchiferan constructions are easily derived from lateral grove forming primitive conchiferans. In every single group the dorsal molluscan body (the shell-stabilized hump with the inner organs) is well demarcated from the ventrally shifted cephalopodium with the muscular apparatus. By partial reduction of dorsoventral musculature and as a consequence, reduction of gills and a reduction or unification of the nephridia constructions arose, which seem to be non-metameric and non-coelomatic.

A spiral coiling of the former cup-shaped shell and a reduction of most of the dorsoventral muscles, leaving only one crossing pair at least, caused torsion and lead to the gastropod construction (Fig. 4) (Edlinger 1988a, b, 1989a, 1991a). This construction gave rise to a radiation of many different groups varying in a high degree. But nevertheless all these different types follow out of a common ancestral construction which can be reconstructed very precisely.

By a coiling of the shell and a subsequent chambering new, pelagic constructions arose, using buoyancy of a gas-filling. A reduction of some dorsoventral muscles was attended by the step by step-reduction of mantle cavity-organs (Fig. 5) (Bonik, Grasshoff, Gutmann & Klein-Rödder 1977). A dorsal elongation, accompanied by reductions of musculature and also of mantle-cavity organs lead to the scaphopod-construction, which lives digging within soft substrates (Fig. 6) (Edlinger 1991 b). In a way this construction represents a dorsoventrally extremely elongated monoplacophoran construction.

Neopilina represents a highly specialized monoplacophoran construction. Flattening enabled the filter-feeding *Neopilina* to turn its body into an anatomically inverted position (Fig. 7).

A median subdivision, attended by a rearrangement and a partial reduction of the dorsoventral musculature resulted the bivalves (Fig. 8) (Vogel & Gutmann 1980). As a consequence of these rearrangements and of the change of shell-form and -construction the foot was elongated and changed to a digging organ. The head structures, the radula in particular, must disappear.

The phylogenetical models, outlined above, follow constructional considerations and clear explanations based on law-like concepts. Such a procedure should be mandatory for all phylogenetical reconstructions. In contrast to other phylogenetic considerations, which are inconclusively based only on morphological comparisons or the unregulated use of fossil remains these reconstructions can be strictly tested and criticized out by referring to valid laws and constructional principles as muscle antagonism and form enforcing mechanisms.

The alternative unexplained views

In the literature a great number of considerations for mollusk evolution have been laid down. But only "gestalt" and form sequences can be found. None of the hypotheses is based on explanations. The elucidated constructional principles of the hydraulic concept and constructional morphology have been consistently neglected. The explanatory value of the hydraulic principles and the constructional theory were strictly ignored. It is no exaggeration to diagnose a boycott which must be considered an attempt to keep morphology anchored in the anachronistic tents of traditional gestalt morphology. Some older ideas about mollusk evolution should be mentioned.

Several authors suppose turbellarian flatworms to be the ancestors of mollusks. On the basis of comparisons between Caudofoveata and turbellarians Salvini-Plawen (1968, 1972, 1981) and Salvini-Plawen & Steiner (1996) suggest a turbellarian-like construction with a posterior anus to be the ancestor of the mollusks. Salvini-Plawen & Bartolomaeus characterize the mollusks as "mesenchymata with a coelom". Haas (1981) agrees virtually with Salvini-Plawen. His concept of molluscan phylogeny with aplacophorans as ancestral mollusks is based on a comparison of their hard-parts, the spicules in particular. An origin of mollusks from turbellarians is accepted implicitly. Morphological changes are not outlined. Scheltema (1978) argues more consistently. At the one hand, she emphasizes the similarity of Caudofoveata and Ventroplicata and their possible derived status. At the other hand, she agrees with the theory of turbellarian-like ancestors ignoring totally the difficulties of this hypothesis outlined above. The same is true for her characterization of the internal anatomy of the aculiferous Aplacophora as "primitive". One must say that in this idea the flattening of the body is explainable because the flatworms have a dense tethering of dorsoventral muscles which allow the creeping on the substrate. However, no transition to the radula, the foot, the intestinal canal, the pericardium is conceivable or can be given. Any kind of justification is lacking. Only such ancestors which had a coelom as a mechanically closed hydraulic subsystem around an intestinal canal, which they could shift to the dorsal part of the body, were candidates as mollusk ancestors. The step by step shifting process resulted in an enlargement and a rearrangement of the ventral musculature. For flatworms and nemertineans such a constructional change is not possible.

The idea of a close phylogenetic relation between mollusks and Kamptozoa, presented by Haszprunar (1996) is based on comparisons and cladistic evaluations of various characters, which are conceived as isolated objects without interconnections. No consistent reconstruction of the common ancestor is given. At the other hand his characterization of the mollusks as coelomate but not eucoelomate may be seen as a first step towards a general revision of the turbellarian hypothesis. The same is true for the suggestions of mollusks as "mesenchymata with a coelom" (Salvini-Plawen & Bartolomaeus 1994). Even non-metameric worms were taken into consideration and claimed as possible ancestors. Here we find evidence for a complete non-understanding of form-determination and of constraints in the transition to constructional solutions. Flattening of worms with large coelom chambers without inner tethering structures such as metamerical structures is not possible because the cross-section is restricted to circularity and cannot be changed into a flattened form. Worms without inner tethering structures will by necessity assume a circular cross-section. Transition to flattening is not possible because the early and the intermediate stages would be functionally senseless and apparently functionally not feasible. So it is inevitable that non-metameric systems must be strictly ruled out as mollusk predecessors.

The differentiation and formation of the molluskan body to a lattice-like muscular locomotor organ and to a dorsal part with the inner organs under the vault of the shell as a stabilizing frame is not conceivable from an non-metameric forerunner construction. Only the presupposition of annelid-like ancestors will allow the transformation of the constructions in the adequate way.

It is very astonishing that almost all worm-like organisms have at some time or other been considered candidates as mollusk forerunners. No explanation for any of the implied transitions was given in any one of the papers in the literature. In contrast to other authors arguing for turbellarian-like ancestors morphologically, Götting (1980a, b) tries to give reasons for an annelid-descent of mollusks by morphological and ultrastructural features, but constructional properties by him are treated marginally too.

In the model presented above the status of *Neopilina* as a very specialized construction is emphasized. This might concern also other extant monoplacophorans. So it is very astonishing that Haszprunar (1992a) in accordance to his other derivations (Haszprunar 1988, 1992b) purposes to put some monoplacophorans with diminutive size and of course reduced numbers of segmental elements in the most important position of monoplacophoran and conchiferan phylogenetic tree, believing these dwarfish organisms to be representatives of ancient constructions, evolving to serial monoplacophorans after the event of origin of mollusks. He neglects the everywhere observable fact that a diminutive size is correlated automatically with a simplification of the construction and with a reduction of serial organs. Extreme diminution can be seen as a blind alley of phylogeny. The same is true for Runnegar's (1996) suggestions of "millimetric-sized ancestral conchiferans."

The phylogenetical reconstructions of Lauterbach (1983) are problematic too. He postulates an origin of serial organ-arrangements by a step by step growing process from the rear to the front of primitive ancestors. Constructive implications of the musculature in particular are mostly ignored. It would be easy to continue the specification of invalid attempts.

As like as the phylogenetic reconstructions of zoologists treating extant organisms, the paleontological models mostly are based on morphological characteristics and arbitrary in a high degree. They don't consider constructional properties of organisms and don't give reasons for the alleys of phylogenetical change too. Their evaluation of characters, used for phylogenetical trees, is subjective because no constructional basis of characters is given. Many fossils consist only of parts of skeletons. Therefore the reconstructions of soft-parts of fossils are impossible in many cases. Furthermore it is unclear if the fossil record contains the real ancestors of extant mollusks and how the latter can be identified. This is true in respect to Cambrian polyplacophorans, which are believed to be ancestors of the extant. Ignoring these presuppositions Yochelson (1966) interpreted conical shell-plates of *Mattheva* as originating from an animal with a two-valved shell. This arbitrary two-valved shell was derived from an univalve. Yochelson's conclusion was that the ancestral mollusks were Monoplacophorans.

In contrast to Yochelson's speculation, Runnegar & Pojeta (1974) are claiming for a Chiton-like construction with 7 or 8 shell-plates. In accordance with Yochelson they suppose a high phylogenetical relevance of *Mattheva*. The argument that conical shell-plates as insertions of muscles logically must be an indication for a complicated and specialized construction is unconsidered.

Nevertheless out of constructional reasons Chiton-like constructions must be seen as the ancestors of all polyplacophorans, monoplacophorans and conchiferans. This does not concern the internal structure of the shells resp. the shell-plates. That means that the articulated connection of Chiton-shell-plates or the number and structure of shell-layers in various mollusks are special problems and of less importance for the molluscan phylogeny in general.

In many cases it is quite unclear if fossils may be conceived as mollusks or not. So the theory of Marek & Yochelson (1976), who believed the hyoliths, animals with long conical shells to be mollusks was rejected by Runnegar et. al. (1975) and Bandel (1983). In contrast to the hyoliths the rostroconchs may be real mollusks. Because of their bivalve-shell many authors believe them to be Pelecypoda. Others established a special class rostroconchs because some characteristics of the shells differ from that of "common Pelecypoda". If the reconstruction results a mussel-like construction we must postulate the same phylogenetical changes as suggested by Vogel & Gutmann (1980) for Pelecypods. Only a reconstruction of the rostroconchan soft bodies could answer all questions.

Concerning a morphological trend to shell-elongation among Late-Cambrian and Ordovician rostroconchs, Pojeta Jr. (1987) believed the scaphopods to be rostroconch descendents, but he does not give reasons for all constructional changes, including the changes of musculature which are necessary for the evolution of animals with a tube-like, univalve shell out of bivalve constructions. So it seems to be the only consistent derivation to suggest monoplacophoran-like ancestors of the Scaphopods.

A phylogenetical derivation of the cephalopods was given by Bonik, Grasshoff, Gutmann & Klein-Rödder (1977). In general one can agree with this very detailed model, because it is consequently based on constructional conditions. They avoid consequently identifying fossils with the ancestors of cephalopods. Fossils, presented as possible ancestors of the cephalopods by several authors (e.g. Yochelson, Flower & Webers 1973), can only be taken as supporting evidences, but not as an empirical basis of phylogenetic models. The constructional implications presented above would be even valid if one suggests a polyphyletic origin of some groups of cephalopods.

The derivation of the gastropods from monoplacophoran-like ancestors by a step by step reduction of most dorsoventral muscle-bundles is almost unquestioned. This agrees with Ghiselin's (1966) claim for a continuous series of transitional stages. Torsion is the last consequence of these reduction processes (Edlinger 1988a, b, 1989a).

A misunderstanding of the constructional implications of the monoplacophoran "bauplan" has lead to a phylogenetical interpretation of the patellaceans as connecting link between monoplacophorans and gastropods by Haszprunar (1988a). This model suggests a (logically) saltatorian change from monoplacophorans to patellacean-like gastropods.

Haszprunar does not consider the impressive constructional differences of all organ systems including the musculature. Furthermore he gives no reasons for torsion. Many authors explain torsion by neoteny or by needs of the larva (Garstang 1928, Minichew & Starobogatow 1972). In contrast to these authors we must argue, that larva are ephemeral transitional stages. It cannot be believed that the needs of only few days should have an after-effect for the whole of the life.

In the same way, models must be rejected, which reconstruct the phylogenetical process of torsion in accordance with the biogenetic law on the basis of embryogenesis. Besides the fact that the biogenetic law is based on a vicious circle (Edlinger 1986) it can be shown that the special modes of torsion depend on various conditions of the larval constructions, on the quantity of ooplasm in particular (Tardy 1970).

In the same way as all constructional and morphological subsystems can only be seen as restrained and constrained by the frame construction as a whole, also the histological structures and the fine-structure must be considered as results of constructional properties of the living organisms. All attempts to employ traditional morphological comparative methods at the submicroscopical level and to reconstruct molluscan phylogeny by ultrastructural results without any respect to their construction, as done by Haszprunar (1984a, b) in many cases, must be rejected. Also these attempts must result in valid models.

In the present time many scientists pin their hope on DNA-analyses as most efficient methods for an elucidation of phylogenetic relations, but the results of DNA-analyses in many cases depend on the special sequences, which are studied (Ghiselin 1988, Wägele 1994a,b; Wägele & Wetzel 1994). At the other hand relations and different similarities of DNA-sequences cannot give us any information about the construction of common ancestors and the intermediate states of evolution, that means their anatomy and morphology too. So we can conclude, that there's no getting around phylogentic reconstructions on the basis of the theory of organismic constructions.

Literature

Bandel, K. (1983): Wandel der Vorstellungen von der Frühevolution der Mollusken, besonders der Gastropoda und Cephalopoda. Paläont. Z. **57** (3/4): 271–284.

Bereiter-Hahn, J. (1991): Cytomechanics and biochemistry. – In: Schmidt-Kittler N. & K. Vogel (Eds.): Constructional morphology and evolution. – Springer Berlin-Heidelberg-New York. p.81–90.

Bonik, K., M. Grasshoff, W. F Gutmann und R. Klein-Rödder (1977): Die Evolution der Tintenfische. Ein Entwurf für das Schaumuseum. – Natur u. Museum **107** (8): 244–250.

Edlinger, K. (1986): Hat das "Biogenetische Grundgesetz" ausgedient? – Mitt. Inst. f. Wiss. u. Kunst 2/ 1986: 59–63. Wien.

Edlinder, K. (1988a): Beiträge zur Torsion und Frühevolution der Gastropoden. – Z. zool. Syst. Evolut.-forsch **26**: 27–50.

Edlinger, K. (1988b): Torsion in Gastropods: A phylogenetical Model. – Malacological Rev. Suppl. **4**: 239–248.

Edlinger, K. (1989a): Zur Evolution der Schneckenkonstruktion 1.– Torsion und Frühevolution der Prosobranchier. – Natur u. Museum **119** (9): 272–293, Frankfurt.

Edlinger, K. (1989b): Sekundäre Wurmkonstruktionen bei amphineuren Mollusken. – In: Edlinger, K. (Hgb.): Form und Funktion – Ihre stammesgeschichtlichen Grundlagen. Wiener Universitätsverlag, S. 35–54.

Edlinger, K. (1991a): The mechanical constraints in mollusc constructions - the function of the shell, the musculature, and the connective tissue. – In: Schmidt-Kittler N. & K. Vogel (Eds.): Constructional morphology and evolution. – Springer, Berlin-Heidelberg-New York, p. 359–374.

Edlinger, K. (1991b): Die Evolution der Scaphopodenkonstruktion. – Natur u. Museum. **121** (4): 116-122.

Edlinger, K. (1991c): Die "Unsichtbare Morphologie" - Energiefluß und Konstruktion. – In Gutmann, W. F. (Ed.): Die Konstruktion der Organismen I. Kohärenz, Energie und simultane Kausalität. Aufs. u. Reden Senckenb. Naturf. Ges. **39**: 27–49.

Edlinger, K. (1991d): Nervensysteme als integrale Bestandteile der mechanischen Konstruktion. – Aufs. u. Reden Senckenb. Naturf. Ges. **39**: 131–155.

Edlinger, K. (1994a): Morphologische Determinanten bei der Bildung von Nervensystemen. – Biol. Zentralblatt, **113**: 25–32.

Edlinger, K. (1994b): Das Spiel der Moleküle – Reicht das Organismusverständnis des molekularbiologischen Reduktionismus? – Natur u. Museum **124** (6): 199–206.

Edlinger, K. (1994c): Ontogenetische Mechanismen in Beziehung zur Evolution. In: Gutmann, W. F., D. Mollenhauer und D. S. Peters (Ed.): Morphologie & Evolution. W. Kramer, Frankfurt/ M., S. 365–384.

Edlinger, K. (1995a): Organismen: Gen arrangements oder Konstruktionen? – Eine Replik auf E. Mayr und die Synthetische Theorie. Biol. Zentralblatt.

Edlinger, K. (1995b): Elemente einer konstruktivistischen Begründung der Organismuslehre. – In: Gutmann, W. F. & M. Weingarten (Ed.):Die Konstruktion der Organismen I. Struktur und Funktion. Aufs. u. Reden Senckenb. Naturf. Ges. **43**: 87–103.

Edlinger, K. & W.F. Gutmann (1996): Mollusks as evolving constructions: necessary aspects for a discussion of their phylogeny. – Iberus **15** (2): 51–66.

Edlinger, K., W.F. Gutmann & M. Weingarten (1991): Evolution ohne Anpassung, Verl. W.Kramer, Frankfurt am Main.

Edlinger, K. (1995): Die Evolution der Plathelmintenkonstruktion - Zur Stammesgeschichte und Systematik der Plattwürmer. – Natur u. Museum **125** (10): 305–320.

Garstang, W. (1928): The origin and evolution of larval forms. Brit. Ass. Rep. Glasgow, p. 77–98.

Ghiselin, M.T. (1966): The adaptive significance of gastropod torsion. Evolution **20**: 337–348.

Ghiselin, M.T. (1988): The origin of mollusks in the light of molecular evidence. Oxford Surveys of Evolutionary Biology **5**: 66–95.

Götting, K.J. (1980a): Origin and relationships of Mollusca. – Z. zool. Syst. Evolutionsforsch. **18**: 24–27.

Götting, K.J. (1980b): Argumente für die Deszendenz der Mollusken von metameren Antezedenten. – Zool. Jahrb. Anat. **103**: 211–218.

Gutmann, W.F. (1974): Die Evolution der Mollusken-Konstruktion: Ein phylogenetisches Modell. – Auf. Reden Senckenb. Naturf. Ges. **25**: 1–84.

Gutmann, W. F. (1988) The hydraulic principle. – Am. Zool. **28**: 257–266.

Gutmann, W. F. (1989): Die Evolution hydraulischer Konstruktionen. Organismische Wandlung statt altdawinistischer Anpassung. – W. Kramer, Fankfurt am Main, 201 p.

Gutmann, W.F.& K. Edlinger (1991a): Die Biosphäre als Megamaschine – Ökologische und

paläo-ökologische Perspektiven des Konstruktionsverständnisses der Organismen I. Natur u.
Museum **121** (10): 302–311.

Gutmann, W.F.& K. Edlinger (1991b): Die Biosphäre als Megamaschine – Ökologische und
paläo-ökologische Perspektiven des Konstruktionsverständnisses der Organismen II. Natur u.
Museum **121** (12): 401–410.

Gutmann, W.F.& K. Edlinger (1994a): Organismus und Evolution - Naturphilosophische Grundla-
gen des Prozeßverständnisses. – In: Bien, G. u. J. Wilke (Hrsg.): Natur im Umbruch – Zur
Diskussion des Naturbegriffs in Philosophie, Naturwissenschaft und Kunsttheorie. –
Frommann-Holzboog, S. 109–140.

Gutmann, W.F. & K. Edlinger (1994b): Molekulare Mechanismen in kohärenten Konstruktionen. –
In: W. Maier u. Th. Zoglauer (Hrsg.):Technomorphe Organismuskonzepte – Modellübertra-
gungen zwischen Biologie und Technik. Frommann-Holzboog, S. 174–198.

Gutmann, W.F. & K. Edlinger (1994c): Morphodynamik und Maschinentheorie: Die Grundlage einer kausalen Morphologie. – In: Gutmann W F., D. Mollenhauer und D. S. Peters: Morphologie
and Evolution. W. Kramer, Frankfurt am Main. S. 177–199.

Gutmann, W.F. & K. Edlinger (1994d): Neues Evolutionsdenken: Die Abkoppelung der Lebensentwicklung von der Erdgeschichte. – Archaeopteryx **12**: 1–24. München.

Haas, W. (1981): Evolution of calcareous hardparts in primitive mollusks. Malacologia **21** (1–2):
403–418.

Haszprunar, G. (1984a): The fine morphology of the osphradial sense organs of the mollusca I.
Gastropoda, Prosobranchia. – Phil. Trans. R. Soc. London B **307**: 457–496.

Haszprunar, G. (1984b): The fine morphology of the osphradial sense organs of the mollusca II.
Allogastropoda (Architectonicida, Pyramidlellida). – Phil. Trans. R. Soc. London B **307**: 497–
505.

Haszprunar, G. (1988a): On the origin and evolution of major gastropod groups, with special reference to the streptoneura. Journal of Molluscan Studies **54**: 367–441.

Haszprunar, G. (1988b): Die Torsion der Gastropoda – ein biomechanischer Prozeß? Zeitschrift für
zoologische Systematik und Evolutionsforschung **27**:1–7.

Haszprunar, G. (1992a): The first mollusks: small animals. Bolletino Zoologico **59**: 1–16.

Haszprunar G. (1992b): Preliminary anatomical data on a new neopilinid (Monoplacophora) from
Antarctic waters. Abstracts of the 11th. International Malacological Congress Siena, p. 307–
308.

Haszprunar G. (1996): The mollusca: Coelomate turbellarians or mesenchymate annelids? – In: Taylor J. (Ed.): Origin and evolutionary radiation of the Mollusca. Oxford University Press, p. 1–
28.

Lauterbach, K.E. (1983): Erörterungen zur Stammesgeschichte der Mollusca, insbesondere der Conchifera. – Z. zool. Syst. Evolut. forsch. **21** (3): 201–216.

Marek, L, & E.L. Yochelson (1976): Aspects of biology of Hyolitha (Mollusca). Lethaia **9**: 65–82.

Minichew, W. & J.I. Starobogatow (1972): The problem of torsion process and the pro-morphological
change of the larves of trochophora-animals (in Russian). Zool. Zhurn. Leningrad **11**: 1437–
1141.

Pojeta, J. Jr. (1987): Rostroconcha/Scaphopoda. In: Cheetham, A. H. & Rowell, A. J. (Ed.). Fossil
Invertebrates. Blackwell Scient. Publ. Palo Alto / Oxford /London /Edinburgh /Boston /Mel-
bourne. p. 358–386.

Runnegar, B. (1996): Early evolution of the Mollusca: The fossil record. – In: Taylor J. (Ed.): Origin
and evolutionary radiation of the Mollusca. Oxford University Press.

Runnegar, B. & J. Pojeta (1974): Molluscan Phylogeny: The Paleontological Viewpoint. Science
186: 311–317.

Runnegar, B., J. JR. Pojeta, J. Noel, J. E. Taylor, M.F. Taylor, & G. Mcclung (1975): Biology of the
Hyolitha. Lethaia **8**: 181–191.

Salvini-Plawen, L. (1968): Die 'Funktions-Coelomtheorie' in der Evolution der Mollusken. – Syst.
Zool. **17**: 192–208.

Salvini-Plawen, L. (1972): Zur Morphologie und Phylogenie der Mollusken: die Beziehungen der

Caudofoveata und Solenogastres als Aculifera, als Mollusca und Spiralia. – Z. wiss. Zool. **184**: 205–394.

Salvini-Plawen, L. (1981): On the origin and evolution of the Mollusca. – Atti Convegni dei Lincei **49**: 235–293.

Salvini-Plawen, L & T. Bartolomaeus (1994): Mollusca: Mesenchymata with a "coelom". In: Lanza-vecchhia, G., R. Valvassori & M.D. Candia Carnevale (eds.): Body cavities: function and phy-logeny. – Mucchi Editore, Modena, p. 75–92.

Salvini-Plawen, L. & G. Steiner (1996): Synapomorphies and plesiomorphies in higher classifica-tion of Mollusca. – In: Taylor, J. (Ed.): Origin and evolutionary radiation of the Mollusca. Oxford University Press, p. 29–51.

Scheltema, A.H. (1978): Position of the class Aplacophora in the phylum Mollusca. Malacologia **17** (1): 99–109.

Scheltema, A.H. (1978): Phylogentic position of Sipuncula, Mollusca and the progenetic Aplaco-phora. – In: Taylor, J. (Ed.): Origin and evolutionary radiation of the Mollusca. Oxford Univer-sity Press, p. 53–58.

Tardy, J. (1970): Contribution a l'etude des metamorphoses chez le Nudibranches. Ann. des sci. nat. Zool. Paris:: 299–370.

Vagyvolyi, J. (1967): On the origin of mollusks, the coelom, and coelomatic segmentation. – Syst. Zool. **16**, 153–168.

Vogel, K. & W.F. Gutmann (1980): The derivation of pelecypods: role of biomechanics physiology and environment. – Lethaia **13**: 255–264.

Wägele, J.W. (1994b): Review on methodological problems of "computer cladistics" exemplified with a case study on isopod phylogny (Crustacea:Isopoda). – Z. zool. Syst. Evolut. forsch. (in press).

Wägele, J.W. (1994b): Rekonstruktion der Phylogenese mit DNA-Sequenzen: Anspruch und Wirk-lichkeit. – Natur u. Museum **124** (7): 225–232.

Wägele, J.W. & R. Wetzel(1994): Nucleic acid sequence data are not per se reliable for inference of phylogenies. – J. Nat. hist. **94**: 749–761.

Woltzow, J. (1988): The organization of limpet pedal musculature and its evolutionary implications for the gastropoda. – Malacol. Rev. Suppl. **4**: 273–283.

Yochelson, E.L. (1966): *Matthewa*, a proposed new class of mollusks. Prof. Pap. U.S. Geol. Surv. 532B, 1–11.

Yochelson E.L., Flower, R.H. & Webers, G.F. (1973): The bearing of the new Late Cambrian genus Knightoconus (Mollusca: Monoplacophora) upon the origin of the Cephalopoda. Lethaia **6**: 275–310.

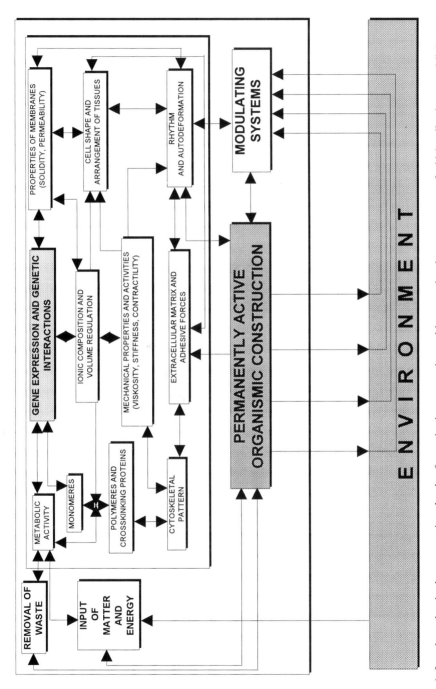

Fig. 1: Interdependencies between various levels of organismic construction and between the elementary parts, of which cells, tissues and living beings as wholes consist.

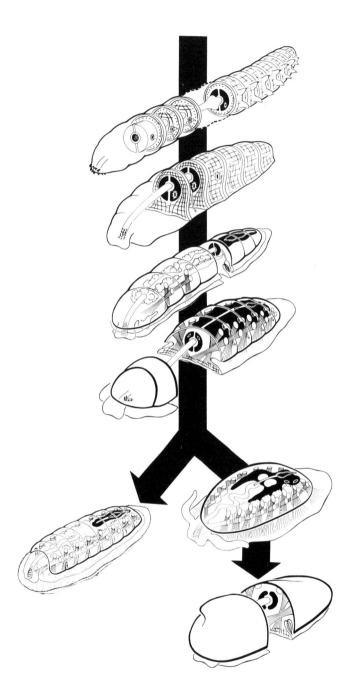

Fig. 2: The evolution of an early mollusc construction out of an annelid ancestor. Out of an early chiton-like ancestor there evolve the amphineurans at the one hand (left branch) and early mono-placophorans (right branch) as the ancestors of the conchiferans at the other.

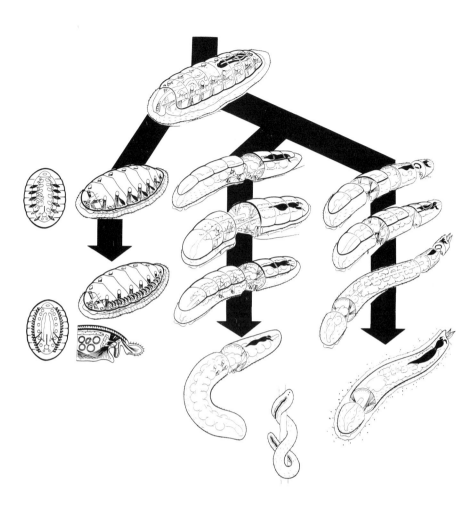

Fig. 3: The evolution of the amphineuran mollusks out of a primitive high chambered chiton-like construction. Left row: The evolution of the recent flattened chitons (Placophora = Polyplacophora). Middle row: The evolution of the Ventroplicata. Right row: The evolution of the Caudofoveata.

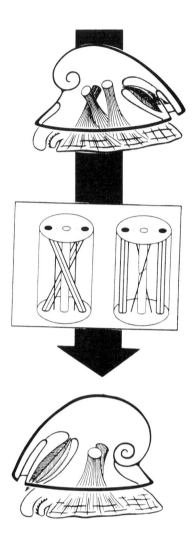

Fig. 4: The evolution of the gastropods by reduction of the dorsoventral musculature and torsion.

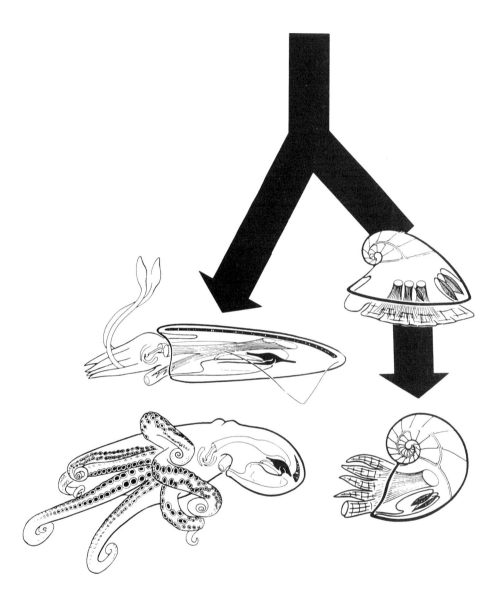

Fig. 5: The evolution of the cephalopods.

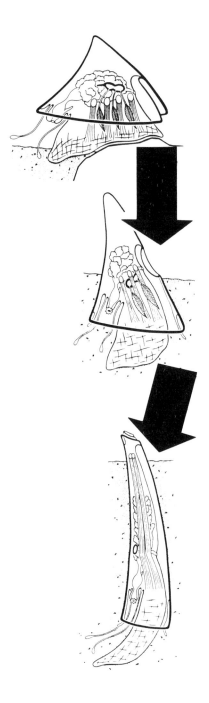

Fig. 6: The evolution of the scaphopods.

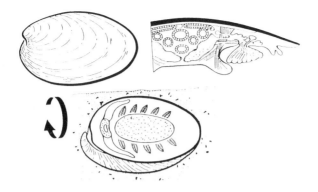

Fig. 7: *Neopilina* (see text).

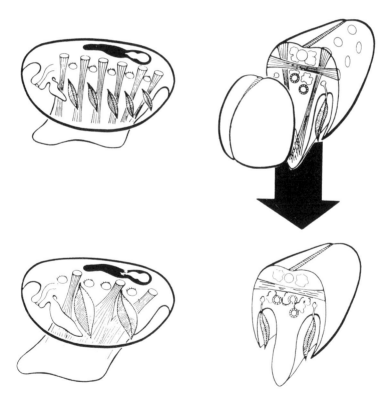

Fig. 8: The evolution of bioalves.

A structural-functional approach to the soft bodies of rugose corals

Michael Gudo

1. Introduction

The study of the generation and control of shape is an integral part of a constructional approach to organisms. If living beings are understood as organismic constructions, the determination of shape can be explained as being driven by hydraulic, mechanical and energy conducing entities. The recent Anthozoans such as the Actiniaria, Octocorallia, or Scleractinia, can be analyzed in such a manner that it leads to an understanding of their organisation and the physical laws underlying it.

The most important principles that have to be observed for a constructional approach of such organisms are hydrostatic and hydrodynamic physical laws, the principles of coherence, energy transformation and the mechanisms of muscular action, including the antagonistic interaction necessary for the proper functioning of all contractile elements (Gutmann 1991). A proper understanding of these principles leads to the conclusion that no structural transformation in individual development or evolution is arbitrary or accidental (Gutmann & Bonik 1981, Gutmann 1995, Grasshoff 1976, Peters & Gutmann 1971, Schmidt-Kittler & Vogel 1991).

The results of a constructional morphology of recent organisms can be applied to an analysis of the structural, organizational and functional aspects of fossil organisms. Any anatomical reconstruction of fossil and extinct organisms can only be performed in analogy to that of recent relatives and has to refer to biological and morphological knowledge. In this approach constructional morphology is used as a principle of uniformitarianism (Gudo 1997, Gutmann 1997) in order to reconstruct the lost soft bodies of the fossil rugose corals.

2. The rugose corals and the scleractinian corals

Paleozoic corals that can be considered to be related to the Rugosa with some degree of certainty appeared in the early middle Ordovician and became extinct at the end of the Permian. The scleractinian corals arose during the upper Triassic. The Rugosa flourished mainly in shallow-water environments, in which also calcareous or lime stone reefs were formed, but they were rarely the most important reef builders (Hill 1981). The origin of the rugose corals is uncertain; the relationship between the Rugosa and Scleractinia in uncertain as well, although the Rugosa are classified as a subclass of the Anthozoa. The skeletons of both types of corals share some similarities in their general features, such as the septa or the shape of the corallum. Therefore, the non-fossilized soft parts of the Rugosa are usually considered to show close affinity to those of scleractinian polyps.

Recent scleractian corals consist of a soft body (i.e., polyp) and an exoskeleton. The exoskeleton is secreted by the external epithelium. It consists of small carbonate walls (i.e., septa) and, frequently an epitheca. The main difference between the Rugosa and Scleractinia is the symmetry of the arrangement of septa in their skeletons. The patterns of septa of the Rugosa are bilaterally symmetrical and those of the Scleractinia are radially symmetrical. The Scleractinia insert new septa during individual development in an exponential manner (one new septum into the space between two existing septa, (Fig. 1a). The Rugosa insert new septa in a serial manner (new septa only in four sectors, (Fig. 1b). In their youngest stage the Scleractinia have six septa and spaces between these septa. Afterwards in every space a new septum is inserted. This results in a number of twelve septa and spaces between them (Fig. 1 a).

The Rugosa are characterized by a quite different mechanism of septa insertion. During growth new septa are inserted on one side of the calyx from lateral to medial and, therefore this side is called counter side. On the other side of the calyx new septa are inserted from medial to lateral and, therefore this side is called cardinal side (Fig. 1 b). In many cases the septa have a short radial dimension, so that a calyx is formed.

The skeletons of the Rugosa and Scleractinia show many variations from the septal-insertion patterns shown in Fig. 1. In particular, this pattern of septa and their insertion may not be found in all rugose corals, but it represents a generalization on the basis of which many Rugosa can be explained (probably with the help of relatively simple modifications). However, this assertion needs to be proved for each individual case. In a structural-functional approach, it is important to focus on general constructional and functional aspects in order to explain the differences and variations that can be found in the fossil record.

3. Structural-functional aspects of the soft body of a coral

All organisms generate and preserve their shape by consuming matter and energy. The anthozoan polyps are quite good examples for organisms that determine their shape by the arrangement of tethering elements (Verspannungen) – such as muscles and connective tissue sheets – working against an internal hydraulic fluid. Thereby these organisms generate a pressure on the internal liquid and hence the liquid is a hydraulic skeleton (Chapman 1958).

3.1 Control of body shape and individual growth

The general shape of an anthozoan polyp is barrel-like. Their oral discs are distended and their tentacles have a tendency to round out. The barrel-like shape is formed by three separate muscular systems, namely the circular muscles on the body wall and the radial arranged mesenteries which are composed of vertical and radial oriented muscle fibres. All these muscle fibres attach directly to connective tissue sheets (i.e., the mesogloea). The mesenteries are primary structures that function as tethering devices between the oral and pedal disc. Hence the oral disc is flattened relatively to the pedal disc. The mesenteries hold an hose like flexible tube – i. e. the actinopharynx which is used for fluid exchange – in the center of the polyp. But the mesenteries do not work as a tether between the opposite parts of the outer wall of the barrel (Fig. 2). The round diameter of the barrel is shaped automatically due to the internal pressure of the soft body, because there are no tethering elements to suppress the circular shape of the polyp.

Polyps possess tentacles which are used for several tasks, such as a filling reservoir to control the volume of the gastrovascular cavity, or to capture and store food. As well as internal mesenteries and circular muscles maintain the shape of a polyp, the tentacles are shaped and moved by the activity of muscular fibres and hydraulic fluid. From a constructional point of view, tentacles could be formed only on the oral disc at the sites between the mesenteries. The tentacles can be moved by the activity of circular and longitudinal musclefibres (Fig. 3). In several species, the narrow ends of the tentacles are untethered. Because of the internal hydraulic pressure these narrow ends of the tentacles show a tendency to form a sphere. Other species have small holes at the narrow ends of the tentacles which can be closed. If the internal pressure increases, water is expelled through these holes. Therefore, tentacles are pressure regulating elements that play an important role when a polyp retracts very quickly and continuously. Under several circumstances, for example stress, it may not be possible for the polyp to open the actinopharynx against the very high internal hydraulic pressure (see "3.2 Nutrition"). By developing additional pressure controlling elements (i.e. holes or spheres at the ends of the tentacles) it is possible to decrease the internal pressure so that the mesenteries can open the actinopharynx.

To hold and preserve its barrel-like shape an actinian polyp needs to be anchored to a substrate. Only if the pedal disc is anchored by adhesion to the substrate, the oral disc is flattened (Fig. 4). Thereby, the surface of the substrate is copied by the surface of the pedal disc (Fig. 5). The combined substrate and pedal disc represent an integral part of the basal wall of the polyp.

Polyps adhere only to massive substrates. They are not able to adhere on soft substrates, such as silt or mud for instance. Nevertheless, some polyps attach themselves to a sandy substrate by secreting a substance from their pedal disc. The secreted substance binds the grains of the substrate to each other and connects them to the pedal disc. The same secretion then binds the pedal disc of a polyp to a massive substrate.

In many cases polyps produce a »substrate by their own« instead of attaching themselves to an external substrate. In this case, they generate extracorporeal skeletons through an excessive production of calcium carbonate. These skeletons are secreted by the external epithelium of the lower parts of the polyp. Calcareous matter is produced regular from the whole basal disc, but the matter crystallizes only in the spaces between the mesenteries. This leads automatically to the generation of small calcareous walls in the spaces between the mesenteries. These walls are the septa. They are an exact copy of the internal structure of the soft tissues of the polyp.

Calcareous septa and soft tissue mesenteries are like casts (septa) and molds (spaces between the mesenteries) (Fig. 6).

In order to maintain the barrel shape of a polyp, mesenteries are arranged in at least six directions. There are either six pairs of mesenteries or eight single mesenteries. Four single mesenteries or two pairs of them flatten and maintain the function of the actinopharynx (see "3.2 Nutrition"). These are the »directive mesenteries« or »directives«. Either four additional single mesenteries or pairs of mesenteries stabilize the spaces surrounding the actinopharynx. These are the »lateral mesenteries« or the »laterals«. They are able to open the pharynx.

It is obvious that mesenteries have to be arranged in at least six directions of a very small polyp. Less than six tethering mesenteries would result in uncontrolled bulging of the oral disc (Fig. 7). A larger polyp needs more than six tethering elements. Therefore, new mesenteries have to be added when polyps grow.

The directive mesenteries fix the actinopharynx in its position and control its opening, distension and eversion. They are arranged in two pairs in order to maintain the siphonoglyphes for the exchange of fluid. Therefore, the retractor muscles of the directives are located on the non-opposite parts of the mesenteries. If they were present opposite to each other, the bulging of the retractor muscles during retraction would deform the mesenteries, because of the small space between the mesenteries (Fig. 8).

Most living polyps grow during their individual development. If polyps grow, their diameter as well as the spaces beåween the mesenteries increase. To maintain shape and function of the whole construction and to prevent uncontrolled bulging of the body wall or of the oral disc, new tethering elements have to be inserted. Therefore, polyps grow by developing additional mesenteries and increasing their diameter at the same time (Fig. 9). The Scleractinia and Actiniaria have pairs of mesenteries and they develop new pairs of mesenteries in all spaces between the preexisting mesenteries. The result of this is a mesentery insertion in an exponential manner. This pattern of mesentery insertion casts into the pattern of septa insertion in the skeleton. New septa are added in the spaces between the existing septa. Exponential insertion of mesenteries lead to exponential insertion of septa.

3.2 Nutrition

Polyps acquire nutrients and get rid of waste products by exchanging the fluid in their gastrovascular cavity. Exchange of fluid is achieved by an actinopharynx that extends like a flexible and collapsible tube from the mouth opening into the oral disc inside the barrel. In the two narrow parts of the flattened tube a jelly cushion preserves two cilia stripped channels, the siphonoglyphes. Striking of the cilia in the siphonoglyphes convey water through the actinopharynx in the internal cavity. When fluid is expelled before, the conveyed water refills the gastrovascular cavity. Simultaneously the internal pressure of the polyp increases and the actinopharynx is bilaterally flattened and compressed by the action of the internal pressure and by the pull of the directive mesenteries. This mechanism follows the principle of a valve (Fig. 10).

If the shape of the polyp is stabilized by the vertical and circular muscles, contraction of the radial muscles on the lateral mesenteries will open the actinopharynx against the internal pressure. Then the polyp shrinks and the fluid within the gastrovascular cavity is expelled. Any contraction of only one of the radial or longitudinal muscles will result in a change of the body shape. Contraction of the radial muscles would elongate the polyp, contraction of the longitudinal muscles would shrink the polyp (Fig. 11).

4. Reconstruction of rugose corals

The indicated constructional elements and their relation to an operationally closed, coherent soft body organization refers to the recent polyps of the Scleractinia and Actiniaria. Because the regular principles of structure and organization are uniform in the history of earth and life and because septa and mesenteries behave like casts and molds, it is possible to reconstruct the internal structure of the Rugosa.

The fossil relics of the Rugosa show a specific configuration of newly installed sclerosepta during ontogeny. At this time the reconstruction refers to the generalized representation of the septal insertion mechanism which is presented in most common textbooks of fossil organisms (Fig. 12). The reconstruction will be valid only for this general representation, but it should be tested on the diversity of rugose corals.

4.1 Structural functional aspects of the soft body of a rugose coral

In the general representation of the soft body of an anthozoan coral, I defined the most important constructional elements such as hydraulic fluid enclosed in a muscular tethered cover, internal mesenteries with two muscular layers attached to a connective tissue sheet, a collapsible pharynx, a flattened mouth field, tentacles and the substrate as an instrumentalized part of the construction. All these elements are obligatory for a rugose polyp. For a reconstruction of a rugose polyp, the model is constituted in a way that provides an explanation of special features of the rugose skeletons. Some of these special features are the bilateral symmetry and the septa insertion in four sectors of the diameter as shown in Fig. 1 or Fig. 12.

The soft bodies of the Scleractinia grow in an exponential manner, because they have pairs of mesenteries. New pairs of mesenteries are added in the spaces between the already existing mesenteries (Fig. 9). As we know from other recent cnidarians, pairs of mesenteries are not the only way to tether a barrel-like shaped soft body. For example, the octocorals are tethered by single mesenteries, and the hydrozoans have no mesenteries at all but rather their form is stabilized by the connective tissue and an extraepithelial chitin-layer. This leads to the question whether the rugose polyps were tethered by single or by pairs of mesenteries or whether they had no mesenteries, as Birenheide (1965) postulated. There are several aspects speaking against the ideas of Birenheide. The rugose corals have septa and if they have septa they have to have mesenteries, because septa are formed only in spaces between mesenteries.

Another aspect is that in order to control the shape of the soft body of a rugose coral tethering elements are absolutely necessary. Only very small polyps, such as the hydrozoans, can reduce their mesenteries and control their shape only by circular and longitudinal muscles and a stabilizing connective tissue.

It is a general concept in all organisms to save energy by developing structures which permanently stabilize parts of the soft body. The calyx of the Rugosa is such a structure. The massive carbonate wall of the calyx stabilizes the shape of the basal parts of the soft body. Hence all the mesenteries in those parts of the soft body that excrete the calyx are reduced. The result are wedge-like mesenteries and short septa that do not reach far into the center of the calyx. Shorter septa depend on the generation of shorter (wedge-like) mesenteries (Fig. 13). If the mesenteries are completely missing in the center of the soft body, no septa are formed in the center of the calyx. In the case that carbonate excretion by the soft body is maintained disseppiments are formed. Disseppiments are bubble like carbonate structures. The calyx of the rugose corals is part of the construction that gives important and helpful hints where the understanding and reconstruction of their soft bodies in concerned (see Weyer 1972).

4.1.1 Had the Rugosa single mesenteries or pairs of mesenteries?

Usually the result of pairs of mesenteries is a radial symmetry and exponential insertion of new pairs of mesenteries. Septa were formed in the spaces between the pairs of mesenteries. This is the general mechanism of the scleractinian corals. However, the rugose corals show a different symmetry and mechanism of the septa insertion. Thus, the arrangement and insertion of mesenteries should be different.

Pairs of mesenteries which are peculiar in all known Scleractinians or Actinians have been postulated for the Rugosa by Schindewolf (1950), Hyman (1940), Hill (1981), or Oliver (1980). But no one has tried a reconstruction of the individual development of a coherent and continuously working polyp. As we know from the Scleractinia, growing of a polyp by adding pairs of mesenteries would result in a radial symmetry of the mesentery pattern as well as in a radial symmetry of the septal pattern. If the soft bodies of the Rugosa were tethered by pairs of mesenteries, septa would have to be formed in the small spaces between the twins of a pair of mesenteries. In case this arrangement applied, new pairs of mesenteries would have to be added by the pattern that is peculiar for the Rugosa. But the question: »why aren't there any septa in the spaces between the pairs of mesenteries?« would remain unanswered.

There is another argument against pairs of mesenteries in the soft bodies of the Rugosa. In the Rugosa younger septa split of the older ones (Fig. 12). This was possible only when one existing space was divided into two new spaces while new single mesenteries were added in a serial manner. This results in a separation of the septa from each other. If mesenteries are inserted as pairs, and if septa are formed in the small spaces between the twins of a pair of mesenteries, these twins have to be torn apart if a new pair of mesenteries is inserted in such a manner that a new septum splits from an older septum (Fig. 15).

4.1.2 Individual development of the Rugosa

By assuming single mesenteries to tether the soft body of a rugose coral, it is possible to add new mesenteries in the same manner during the entire individual development. In the soft body of the Rugosa there are four sectors in which new mesenteries are added and hence in the exoskeleton there are also four sectors in which new septa are inserted. There are two different modes of installation of new mesenteries during growth. Addition of new mesenteries occurs in every individual on the cardinal side between the directives and the youngest lateral mesenteries and on the counter side on the far side of the youngest lateral mesenteries.

The reconstruction of the ontogeny of the rugose corals starts with a polyp having six protosepta. These protosepta are produced in the spaces between eight mesenteries. There are four directives (directive mesenteries) and four laterals (lateral mesenteries) (Fig. 16). On the cardinal side the first generation of new mesenteries (b) is inserted between the directives and the first laterals (a). The second generation of lateral mesenteries (c) is now inserted next to the directives while the mesenteries of the first generation (b) is shifted from medial to lateral. This mechanism is continued and maintained in all following growth steps. The proto-laterals and all generations of new mesenteries on the cardinal side are shifted laterally while the following mesentery generation is added.

On the counter side the first generation of new mesenteries is also installed in the spaces between the directives and the laterals, but the second generation (2) is inserted next to the laterals and not next to directives as on the cardinal side. This continues unchanged in further growth. The third generation (3) is also inserted next to the laterals, and so on (4). The new mesenteries of the counter side are also shifted from lateral to medial. The first lateral mesenteries of the counter side are shifted from medial to lateral.

Continuation of this mesenterial insertion mechanism will result in a septal and mesenterial pattern shown in Fig. 16. The stages in ontogeny of the rugose skeleton presented here correspond with the reconstructed phases of mesentery insertion in a rugose soft body (see also Gudo 1997a,b,c).

There are several Rugosa known that show a septal symmetry that cannot be explained by single mesenteries. But within the subclass Rugosa there exists a general pattern of septal insertion that can only be explained by single mesenteries. It is possible that some of the rugose corals have had pairs of mesenteries but the model presented here refers to those rugose corals which correspond to the general pattern of septa insertion. Those Rugosa which show a different symmetry have to be reconstructed by other models.

4.1.3 Potential of rugose corals to form reefs

Corals are some of the most important reef builders today and also have been since the upper Triassic. The rugose corals have not formed the massive reef complexes which we know from the scleractinian corals. In the Paleozoic other organisms formed bigger reefs than the rugose corals. This is a very interesting point, because at first sight corals are more efficient reef builders than for instance porifera or bryozoans. The corals are able to utilize the third dimension of the water column by growing on the skeletons of their parents. This leads to the question whether an explanation of this phenomenon could be provided by the reconstruction of the soft bodies of the rugose corals. The question whether there are structural-functional limitations for the ability of rugose corals to form colonies and to form reefs by themselves as e.g. the recent scleractinians do nowaday arises.

The ability of coral polyps to form reefs could be explained by the optional ontogenetic growth. The polyps produce an extracorporeal skeleton which can be used as a substrate for the following generations of polyps. This would certainly be more effective for the production of a massive reef complex than starting on the seafloor again. Nevertheless larvae settlement is a strategy for the invention of new environments and expanding the outspread of a coral. Every generation of coral polyps will produce its own substrate on the top of the substrate of the older generation. The colony will grow and the third dimension of the water column is utilized as environment.

If one soft body divides itself into two or more new soft bodies that also form an exoskeleton, the result is either budding and dendroid growth or meandroid growth. The soft bodies of the Scleractinia are able to divide themselves into two or more new soft bodies without any loss of the coherent organization or function. This capability is based on adding new pairs of mesenteries in the manner of exponential bifurcation. The soft bodies of the Scleractinia can grow in every direction whereby new mesenteries are inserted (Fig. 17).

The reconstructed soft bodies of the Rugosa are not able to grow in any direction, because they have only four insertion zones. Two of these mesentery insertion zones are located near to the directives on the cardinal side and the two other are located near to the proto-laterals on the counter side. Adding of mesenteries in four insertion zones would automatically result in a round diameter of the soft body. Only if mesenteries were added in one of the existing mesentery insertion zones directive growth would be possible, but then the result is a highly disordered and asymmetrical soft body. There is no way to originate new mesentery insertion zones in one enlarged area of a rugose soft body. Rugose corals which were tethered by single mesenteries and which grow only in four zones are never able to divide one polyp into two or more next-generation-polyps. This is a structural-functional limitation that does not allow the rugose polyps to form as massive reefs as we know from the scleractinian corals.

Nevertheless examples of rugose corals are known which display budding of the polypars. I will try to give two explanations for these examples of the Rugosa. On the one hand it could be that the budding of the skeletons is just imaginary. The budding could be a consequence of the forming of two or more new polyps in the older polypar (Fig. 18). On the other hand it might be possible that rugose corals which show this kind of growth cannot be explained by the model presented here. For these examples of the Rugosa we need another model with a different arrangement of mesenteries and a different pattern of adding new mesenteries. There are many different types of corals collected in the subclass Rugosa and probably taxonomy does not represent a common soft body construction but different ones.

5. Acknowledgments

This work contains the main conclusions of my diploma-thesis which was finished in spring 1996 under the supervision of Prof. K. P. Vogel and Prof. W. F. Gutmann. This text is the printed version of a talk and two posters presented on the 7. Senckenberg Conference: Organisms, Genes, and Evolution in 1996. I feel especially obliged to my late supervisor professor Wolfgang Friedrich Gutmann who gave me the chance to hold this talk and present the posters. I thank Prof. K. P. Vogel for his critical comments, and maintaining the discussion on the reconstruction of the soft bodies of the Rugosa.

6. Literature

Birenheide, R. (1965): Haben die rugosen Korallen Mesenterien gehabt? – Senck. leth., **46**: 27 – 34, 5 Abb.; Frankfurt am Main.

Birenheide, R. (1978): Rugose Korallen des Devon. – **2**: 265 S.; Berlin.

Chapman, M. (1958): The hydrostatic skeleton in the Invertebrate – Biol. Rev., **33**: 338 – 371, 4 Abb; Cambridge.

Grasshoff, M. (1976): Das "Konstruktionsniveau" in der Phylogenetik. – In: Schäfer, W. (1976): Evoluierende Systeme I und II. – Aufsätze u. Reden senckenberg. naturf. Ges., **28**: 124 – 140; Frankfurt.

Gudo, M. (1997a): Konstruktionsmorphologische Rekonstruktion Rugoser Korallen – Profil **11**: 325 – 340, 17 Abb; Stuttgart.

Gudo, M. (1997b): Ist die Konstruktionsmorphologie ein Aktualistisches Prinzip der Paläontologie? – In: Hüssner, H. & Betzler, C. (Eds.): Vogel-Festschrift – Cour. Forsch.-Inst. Senckenberg, **201**: 145 – 160, 7 Abb; Frankfurt am Main.

Gudo, M. (1997c): Reconstruction of rugose corals – a constructional approach. – Newsl. fossil Cnid. Porif., 26/1, 6 Abb.; Graz.

Gudo, M. & Hubmann, B. (1997): Fremdkörpereinschlüsse in Paläozoischen Korallen aus Sicht der Konstruktionsmorphologie – Geol. Paläont. Mitt. Innsbruck, **22**: 22–69, 7 Abb.; Innsbruck.

Gutmann, W. F. (1966): Zu Bau und Leistung von Tierkonstruktionen (4–6). – 4. Bindegewebe und Muskulatur im Bewegungsapparat von Metridium senile. – 5. Struktur und Mechanik des Bindegewebes bei Urticina felina und Sagartia troglodytes. – 6. Funktionelle Gesichtspunkte zur Phylogenie der Coelenteraten: Radialsymmetrie und Muskelapparat. – Abh. Senckenberg. naturf. Ges., **510**: 1 – 106; Frankfurt am Main.

Gutmann, W. F. (1991): Organismus und Energie – Ist die Morphologie noch zu retten? – Naturwiss. Rdsch., **44** (7): 253 – 260.

Gutmann, W. F. (1995): Die Evolution hydraulischer Konstruktion – organismische Wandlung statt altdarwinistischer Anpassung. – Senckenberg Buch, 65: 220 S., 2. Auflage.; (Kramer) Frankfurt am Main.

Gutmann, W. F. (1997): Vom Lyellismus und Adaptationismus zur Konstruktions-Autonomie: Eine Studie zur Grundlagenproblematik der Paläontologie – In: Hüssner, H. & Betzler, C. (Eds.): Vogel-Festschrift – Cour. Forsch.-Inst. Senckenberg, **201**: 161 – 176, 4 Abb.; Frankfurt am Main.

Gutmann, W. F. & Bonik, K. (1981): Kritische Evolutionstheorie – Ein Beitrag zur Überwindung altdarwinistischer Dogmen. – 227 S.; (Gerstenberg) Hildesheim.

Hill, D. (1951): The Ordovician corals. – Proc. Roy. Soc Queensland, **62**: 1 – 27; Queensland.

Hill, D. (1981): Rugosa and Tabulata. – In: Moore, R. C., Robinson, R. A. & Teichert, C. (Eds.) (1981): Treatise on Invertebrate Paleontology, part F: Coelenterata, suppl. 1: – 762 S., 462 Abb.; Lawrence.

Hyman, L. H. (1940): The Invertebrates – Protozoa through Ctenophora. – Vol. 1. – 726 S., 221. Fig.;
  New York and London.
Oliver, W. A. Jr. (1980): The relationship of the scleractinian corals to the rugose corals. – Paleo-
  biology, **6**: 2, S. 146–160.; Lancaster.
Peters, D. S. & Gutmann, W. F. (1971): Über die Lesrichtung von Merkmals- und Konstruktions-
  Reihen. – Z. zool. Syst. Evolutionsforsch., **9** (4): 237 – 263; Hamburg.
Schindewolf, O. (1930): Über die Symmetrie-Verhältnisse der Steinkorallen. – Paläont. Z., **12**: 214 –
  263, 60 Abb.; Berlin.
Schindewolf, O. (1939): Stammesgeschichtliche Ergebnisse an Korallen. – Paläont. Z., **21**: 321 –
  340, 4 Abb.; Berlin.
Schindewolf, O. (1950): Grundfragen der Paläontologie. – 506 S., 332 Abb., 32 Taf.; Stuttgart.
Schindewolf, O. (1967): Rugose Korallen ohne Mesenterien. – Senck. leth., **48**: 134 – 145, 7 Abb.;
  Frankfurt am Main.
Schmidt-Kittler, N. & Vogel, K. (Eds.) (1991): Constructional Morphology and Evolution. – 409 S.;
  Heidelberg.
Weyer, D. (1972): Zur Morphologie der Rugosa (Pterocorallia). – Geologie, 21: 6, S. 710–734., 7
  Abb., 2 Taf.; Berlin.

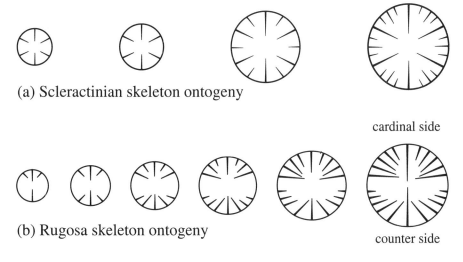

(a) Scleractinian skeleton ontogeny

cardinal side

(b) Rugosa skeleton ontogeny

counter side

Fig. 1: Difference in the skeletons of Rugosa and Scleractinia
Insertion of calcareous septa during the individual development of the Scleractinia (a), and the individual development of the Rugosa (b).

general form:
barrel shape

tethering mesenteries:
tether between upper
and lower side

a hose like tube fixed
by the mesenteries

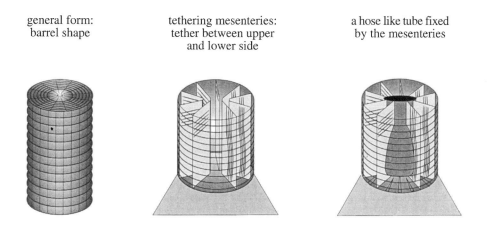

Fig. 2: How to produce a barrel shape
Mesenteries are composed of two muscular layers (vertical and radial) which are directly attached to a connective tissue sheet (mesogloea). The mesenteries generate an internal hydraulic pressure that acts as antagonism and hydraulic skeleton. The mesenteries tether the upper and lower side of the barrel and fix the pharynx in the center of the barrel. Thus the mesenteries are capable of flattening the upper and the lower side as well as to suppress the tendency of the mouth field to round out.

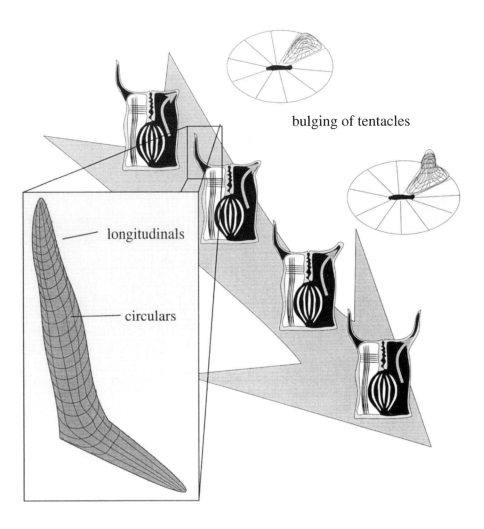

bulging of tentacles

longitudinals

circulars

Fig. 3: Formation of tentacles between mesenteries
The tentacles are hydraulically filled extends of the oral disc of a polyp. They bulge only at the sites
of the mesenteries. The shape of the tentacles is controlled by longitudinal and circular muscles.
They can only be formed between two mesenteries. There are many functions which the tentacles
perform, as for example storage of food control of the hydraulic volume or the internal pressure.

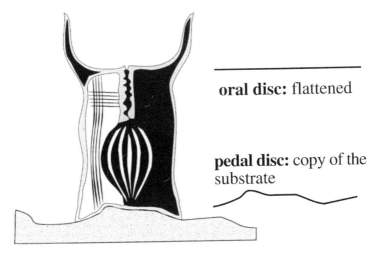

oral disc: flattened

pedal disc: copy of the substrate

Fig. 4: Flattening of the oral disc
An actinian polyp needs to be anchored to a substrate. Then the surface of the substrate is molded by the surface of the pedal disc, and the oral disc is flattened relative to the pedal disc. The combined substrate and pedal disk can be understood as integral part of the wall of the polyp.

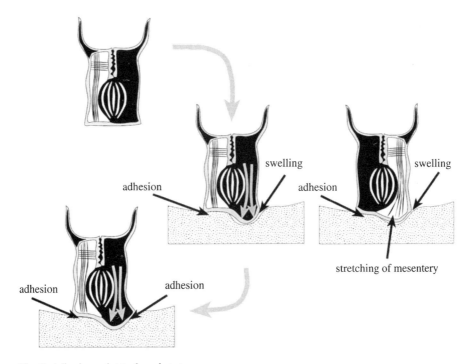

Fig. 5: Adhesion point to the substrate
The substrate becomes part of the construction, and it is a supporting barrier to the hydraulic filling. The self generated internal pressure enforces swelling of all non fixed or tethered parts of the polyp. Thereby the tethering elements will be stretched until the pedal disc is an exact copy of the substrates surface.

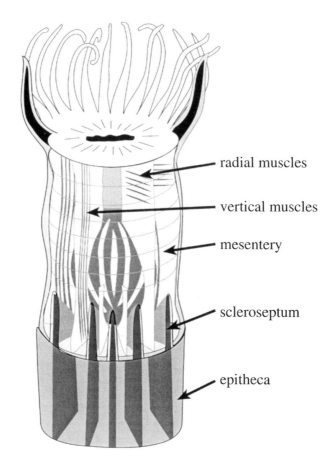

radial muscles

vertical muscles

mesentery

scleroseptum

epitheca

Fig. 6: Extracorporeal skeleton
The exoskeleton is excreted by the external epithelium of the lower parts of the polyp. Calcareous matter is produced regularly from the whole basal disc but crystallization is only possible in the non-movable parts of the basal disc. This leads automatically to the generation of septa in the spaces between the mesenteries. Septa and mesenteries are like casts and molds.

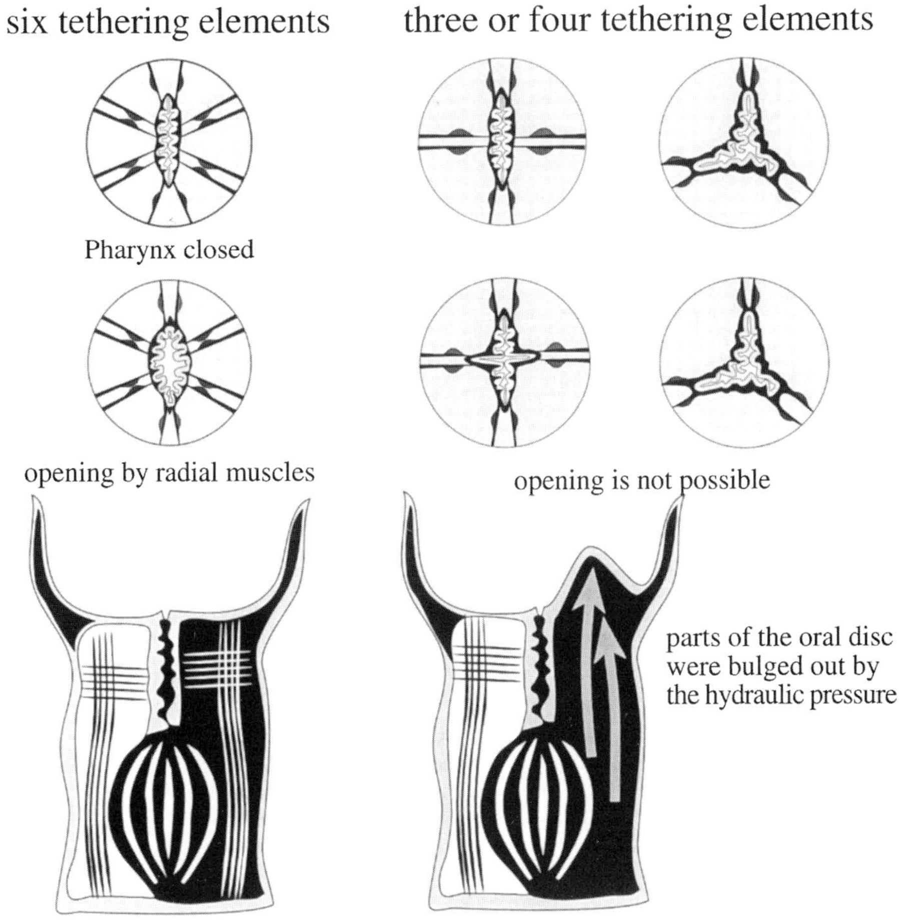

six tethering elements

Pharynx closed

opening by radial muscles

three or four tethering elements

opening is not possible

parts of the oral disc
were bulged out by
the hydraulic pressure

Fig. 7: Minimum number of mesenteries
To preserve the barrel shape, a minimum number of mesenteries is necessary. This number depends on the size of the polyp, but with less than six directions of mesentery tethering the oral disc could not be kept flat at all (right). It does not matter if the six tethering mesenteries are single mesenteries or pairs of mesenteries. It is important that there are directive mesenteries to flatten and to fix the actinopharynx and lateral mesenteries to open the actinopharynx against the hydraulic pressure.

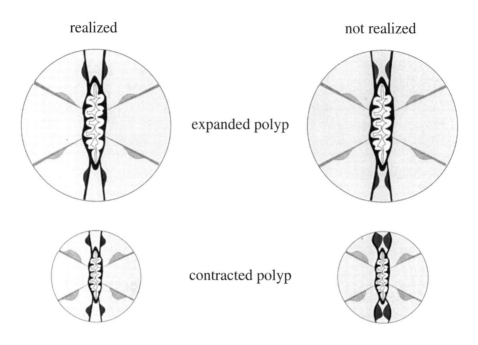

realized   not realized

expanded polyp

contracted polyp

Fig. 8: Arrangement of retractors on the directives
The directive mesenteries are arranged in two pairs in order to allow the maintenance of the sipho-noglyphes. Therefore, the retractor muscles of the directive mesenteries are located on the non-opposite parts of the mesenteries. If they were oriented opposite to each other, the bulging of the retractor muscles would deform the mesenteries, because of the small space between the mesenteries (after Gutmann 1966).

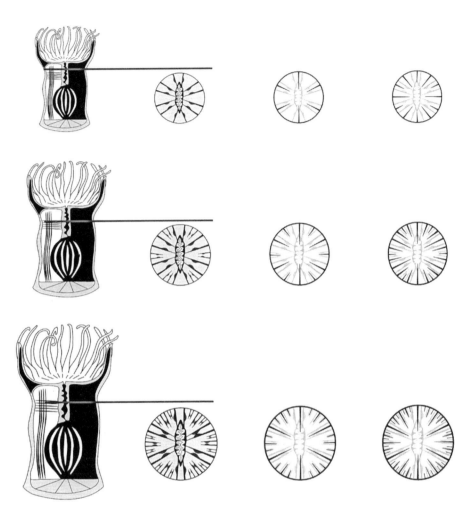

Fig. 9: Growth of polyp constructions
Actiniaria and Scleractinia grow by adding new pairs of mesenteries in the spaces between the existing pairs of mesenteries. When septa are generated in the spaces between the mesenteries, the septa
are added by the same mechanism: a new septum between two existing septa. From the Scleractinia
there are different arrangements of mesenteries and septa known. Septa can be formed within in the
spaces between the pairs of mesenteries, or within the pairs of mesenteries or they are formed as well
within the pairs of mesenteries and within the spaces between the pairs.

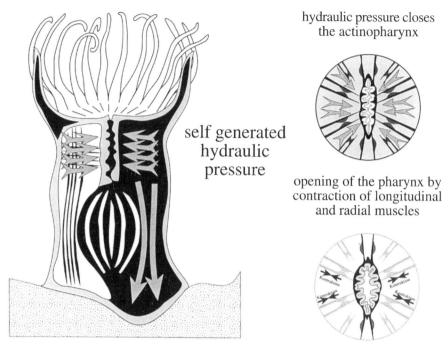

Fig. 10: Valve function of the actinopharynx
The striking of the cilia conveys water into the gastrovascular cavity of the polyp. When fluid was expelled before, this water current refills the gastrovascular cavity. Simultaneously the internal pressure of the polyp increases and the actinopharynx is laterally compressed. This mechanism follows the principle of a valve.

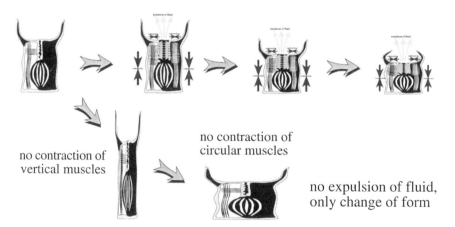

Fig. 11: Expulsion of fluid
Fluid is expelled if the shape of the polyp is stabilized by the vertical and circular muscles. If this is the case, then the actinopharynx can be opened by the radial muscles. If the actinopharynx opens while the longitudinal muscles contract, the polyp shrinks and the fluid within the gastrovascular cavity is expelled. In the case that the shape and size of a polyp are not kept stable, then any contraction of the radial or longitudinal muscles will only result in a change of the body shape.

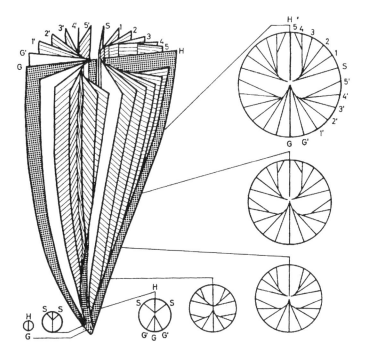

Fig. 12: Generalized representation of rugose coral skeletal growth
From Birenheide (1978).

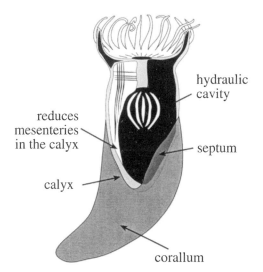

Fig. 13: Rugose corals with reduced mesenteries in the calyx
The massive carbonate wall of the calyx stabilizes the shape of the basal parts of the soft body and
hence all the mesenteries in those parts of the soft body that excrete the calyx can be reduced. The
result are wedge-like mesenteries and short septa that does not reach far into the center of the calyx.
Shorter septa depend on the generation of shorter (wedge-like) mesenteries.

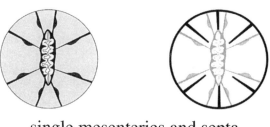

single mesenteries and septa

pairs of mesenteries and septa

Fig. 14: Two kinds of tethering: single or pairs of mesenteries.

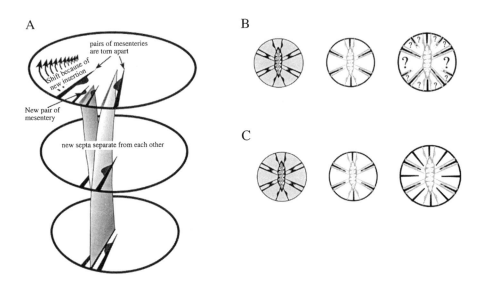

A

pairs of mesenteries
are torn apart

Shift because of
new insertion

New pair of
mesentery

new septa separate from each other

B

C

Fig. 15: Pairs of mesenteries are impossible for the serial septal insertion
The model of Schindewolf (1950) with pairs of mesenteries with the septa in the entocoel of the
mesenteries (=space in one pair) is impossible, because the two singles of one pair would be torn
apart in further development (A). Additionally it is an open question why septa should be generated
only in the entocoel and not also in the exocoel (=space between two pairs) (B). If the main septa are
formed in the exocoel and small septa in the entocoel, a different pattern in generated that does not
correspond to the rugose pattern (C).

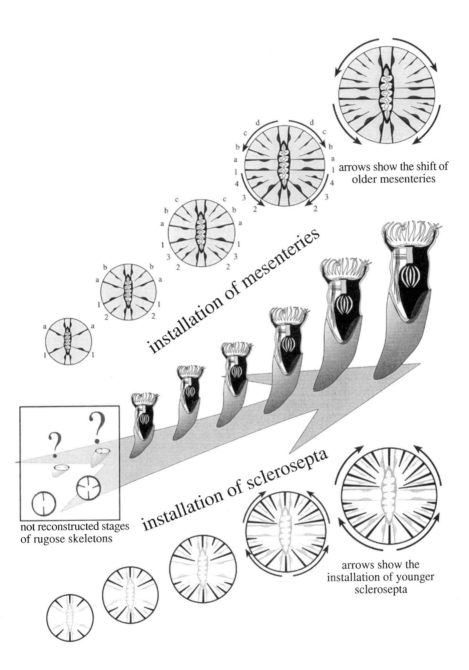

arrows show the shift of
older mesenteries

installation of mesenteries

installation of sclerosepta

not reconstructed stages
of rugose skeletons

arrows show the
installation of younger
sclerosepta

Fig. 16: Insertion of new mesenteries and septa
Model for the addition and integration of new mesenteries and septa in rugose constructions during
enlargement of the polyps. The arrows show the shift of older mesenteries (left hand side) and septa
(right hand side) enforced by the insertion of new mesenteries.

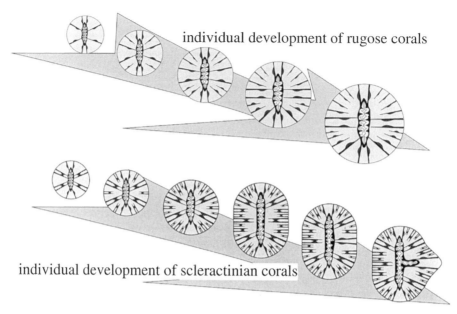

Fig. 17: Differences in growth of rugose and scleractinian polyp-constructions
Comparison of the ontogeny of the rugose polyps as they were reconstructed here and the ontogeny of scleractinian polyps. The Rugosa have only four insertion zones who automatically restrict a round shape of the diameter. Directive growth which is possible for the scleractinians and which is a necessary precondition for budding is not possible if there are only four zones of mesentery insertion. The presented variations of the Rugosa are impossible. First they represent an incoherent organization and second this kind of growth can only happen if there were mesentery insertion zones between the two proto-laterals on each side of the polyp.

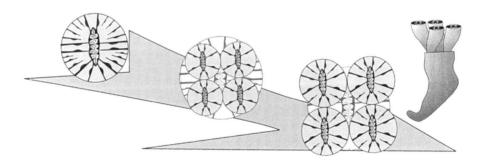

Fig. 18: One possibility of colony generation in the Rugosa
Because no continuous way for dividing a rugose polyp construction into two or more new polyps was found it is proposed that budding of rugose corals is just imaginary. The development of new polyps in the polypar of older ones leads automatically to a kind of budding of the polypars. This is only one explanation for rugose corals that display budding. Another explanation is that the model presented here does not fit for these kinds of corals.

CONTRIBUTORS

Prof. Dr. Walter Bock
Department of Biological Sciences
Columbia University
1200 Amsterdam Avenue
Mail Box 5521
New York, NY 10027 USA

Prof. Dr. Dr. Hans-Rainer Duncker
Institut für Anatomie und Zellbiologie
Justus-Liebig-Universität
Aulweg 123
D - 35385 Giessen

Dr. Karl Edlinger
Naturhistorisches Museum Wien
Burgring 7
A - 1014 Wien

Prof. Dr. Raphael Falk
Department of Genetics
The Hebrew University of Jerusalem
Givat Ram,
Jerusalem 91904, Israel

Dr. Michael Gudo
Neebstr. 11
D - 60385 Frankfurt

Dr. Dr. Mathias Gutmann
Institut für Philosophie
Blitzweg 16
D - 35032 Marburg

Dr. Bernd Herkner
Staatliches Museum für Naturkunde
Erbprinzenstr. 13
D - 76133 Karlsruhe

Dipl. Biol. Christine Hertler
Forschungsinstitut und Naturmuseum Senckenberg
Senckenberganlage 25
D - 60325 Frankfurt

Prof. Dr. Dominique G. Homberger
Department of Biological Sciences
Louisiana State University
508 Life Sciences Building
Baton Rouge LA 70803-1715 USA

Prof. Dr. Peter Janich
Lehrstuhl 1 für Philosophie
Philipps-Universität Marburg
Blitzweg 16
D - 35032 Marburg

Prof. Dr. Christian Kummer SJ
Hochschule für Philosophie
Kaulbachstr. 33
D - 80539 München

Prof. Dr. Antonio Lima-de-Faria
Department of Genetics
Lund University
Sölvegatan 29
S - 22362 Lund, Sweden

Prof. Dr. Sievert Lorenzen
Zoologisches Institut
Christian-Albrechts-Universität
Olshausenstr. 40
D - 24118 Kiel

Prof. Dr. Werner E. G. Müller
Physiologisch-Chemisches Institut
Abt. Angewandte Molekularbiologie
Johann Gutenberg Universität
Duesbergweg 6
D - 55099 Mainz

Dr. Winfried S. Peters
AK Kinematische Zellforschung
Biozentrum der Universität
Marie-Curie-Str. 9
D - 60439 Frankfurt am Main

Prof. Dr. Harald Riedl
Naturhistorisches Museum Wien
Burgring 7
A - 1014 Wien

Prof. Dr. Giuseppe Sermonti
Biology Forum
Rivista di Biologia
Via Annone 6
I - 00199 Roma

Prof. Dr. Franz M. Wuketits
Institut für Wissenschaftstheorie
Universität
Sensengasse 8
A - 1090 Wien